GEOGRAPHY *inside out*

Space, Place, and Society
John Rennie Short, *Series Editor*

GEOGRAPHY

inside out

RICHARD SYMANSKI

KORSKI

With a Foreword by Peter Gould

SYRACUSE UNIVERSITY PRESS

Copyright © 2002 by Syracuse University Press
Syracuse, New York 13244–5160
All Rights Reserved

First Edition 2002
02   03   04   05   06   07   6   5   4   3   2   1

The paper used in this publication meets the minimum requirements of American National
Standard for Information Sciences—Permanence of Paper for Printed Library Materials, ANSI
Z39.48–1984.™∞

Library of Congress Cataloging-in-Publication Data

Symanski, Richard.
Geography inside out / Richard Symanski with a Foreword by Peter
Gould.—1st ed.
p. cm. —(Geography inside out space, place, and society)
Includes bibliographical references (p.).
ISBN 0-8156-0732-6 (cl. : alk. paper)
1. Geography—United States. 2. Geographers—United States. 3.
Symanski, Richard. I. Title. II. Series.
G65 .S96 2002
910—dc21
2002001154

Manufactured in the United States of America

For Tu-tu

and

for Cole

that he will confront stupidity, pretentiousness,
hubris, dishonesty, venality,
and cowardice—wherever found

I thought this account of the *struldbrugs* might be some entertainment to the reader, because it seems to be a little out of the common way, at least I do not remember to have met the like in any book of travels that hath come to my hands; and if I am deceived, my excuse must be, that it is necessary for travellers, who describe the same country, very often to agree in dwelling on the same particulars, without deserving the censure of having borrowed or transcribed from those who wrote before them.

—Jonathan Swift, *Gulliver's Travels*

Richard Symanski is a professor of ecology and evolutionary biology at the University of California at Irvine. Korski is Richard Symanski's soul mate, genetically identical but otherwise another person.

Contents

Illustrations

Foreword

Peter Gould

RICHARD Symanski and his friend Korski are complex characters, difficult to separate sometimes because they are what are generally known as "inseparable friends." I suggest that rather than becoming impatient with this duo, you sit back, relax, and enjoy their writings, some of which will test the "situatedness" and "positionality" of the most convinced and relativistic postmodernist. In brief, there are essays of the utmost seriousness, and others written just for fun.

Rich, for so he is known, blazed what even he would say was a misspent youth, stealing gasoline, breaking into homes, and, by pouring all of his energy into sports, earning dispiritingly low grades in high school. An overbearing father, who had earned his own way since the age of fourteen, was not simply unsympathetic to any form of further education, but downright antagonistic to it. His mother was different: if he attended San Jose State College (he was admitted on probation with several remedial courses to fulfill), then he could live and eat at home, and work to support himself for twenty hours each week as a dishwasher at the local hospital. But what subjects to take? A cousin, said his mother, was an accountant, making a good living, so why not follow in his golden footsteps?

And accounting it was. With a straight-A average at the end of his sophomore year, he passed all portions of the examination to become a certified public accountant except in business law—which he had never taken. Six months and a course later, he had passed all the requirements and needed only two years of practical experience. This necessity he had already started upon, trading his dish washing for positions with two prestigious accounting firms, the second of which, upon Rich's advice, disassociated itself from a client whose accounts and motives appeared to be anything but aboveboard.

So Rich started his professional life as an accountant, and to this day he takes the notions of "accounting for," or being held accountable, with the utmost seriousness. Accountants are the detectives of the business world, and a failure to uncover fraud can have weighty legal and professional consequences. Scrupulously honest accountants, holding to the highest ethical standards, are not always popular. And as a geographer, Rich has not been popular with some of his fellow geographers either, for he asks those writers he questions to explain, to answer, and to debate. Too often, he receives only silences and embarrassed brush-offs.

We have to realize that geography, until relatively recently, has been a small discipline, the establishment was powerful, and until the sixties few dared to raise their voices in public to establish a strong tradition of criticism. Too often, book reviews were bland anodynes or embarrassing hagiomonic pronouncements by acolytes. Some still are. Reviews of journal articles went unsigned, something considered extraordinary in many scientific fields where, again, the idea of being held accountable is considered so obvious that it needs no debate. So Rich's "critical voice" has been muted, and the challenges have often gone unanswered. Puzzled, one asks, "Why?"

But accounting as a lifelong profession was not for him. He longed to discover people and places—he had ridden his bike from California to Maine to experience such things—and decided to take a second major in geography, a disappointing intellectual experience given the quality of instruction at San Jose State at the time. And so, after a year's break in Australia, where again he discovered fraud in the course of supporting himself and his family as a de facto chartered accountant, he applied to, and was accepted by, the graduate program at Syracuse University, receiving his Ph.D. in 1971.

He held junior faculty positions at the University of Cincinnati and the University of Texas at Austin, but the subject matter of his research produced a reaction in the 1970s that appears almost unbelievable today, a reaction that essentially blackballed him from departments of geography and denied him tenure. And here hangs not only a tale, but also a crucially important experience in his life. In 1981, he published *The Immoral Landscape: Female Prostitution in Western Societies*, a thoroughly scholarly study, published by Butterworths of Toronto, a firm whose reputation for scientific publishing is impeccable. Four years later, the book was nominated by the American Sociological Association for its Distinguished Scholar award, and it has since become a minor classic in sociology.

But not in geography. When Butterworths asked the Association of

American Geographers (AAG) for their membership list, a routine request to send out pamphlet announcements, the then executive director, Patricia McWethy, of her own volition, and without consulting anybody, refused, saying, "A number of members would be offended by our participation in the promotion of this publication." This decision was patently impertinent and outrageous, and Rich, with his strong feelings of justice and "being held accountable," was not prepared to take such a decision lying down. Protest followed protest; many of us, who felt the enormity of this decision of censorship, and the implications it might have in the future for stifling free expression and debate, wrote to successive presidents and executive councils but got nowhere. Like a weak dean backing a destructive head of department, the lines of establishment hierarchy had to be maintained at any cost—including fairness, justice, and accountability.

This terrible experience has colored much of Rich's outlook ever since. Married to an internationally distinguished evolutionary biologist, he holds a tenured position in a department of ecology and evolutionary biology in which he teaches ecology and writing to senior biology majors. In 1981, he also published, with John Agnew, *Order and Skepticism,* a book saturated with Karl Popper's concern that anything worthy of the title *theory* must, in principle, be falsifiable, and making the important point that models really have to satisfy only their own internal criteria, so leading easily to tautologies, whereas genuine theories must be able to withstand the possibility of failing to match empirical observations.

As an example, one thinks of the extraordinary experimentalist Madame Wu, who achieved Nobel fame by demonstrating that parity (the left- and right-handedness of spinning electrons) was simply not true, despite the tenacity to which many theoretical physicists of the time clung to the idea. For them, it was "back to the drawing board." But how many times is "theory" invoked in geography, particularly by the post-fill-in-the-blanks, when there is nothing to test, nothing to falsify, when "theory" is too often claimed to provide an air of profundity for the banal? The result is increasingly incestuous tautologies, and we should not wonder that some of Rich's more recent writings have challenged these poseurs!

Three more books were to follow. *Wild Horses and Sacred Cows* was a firsthand, field-based examination of the damage done to federal lands by the explosive proliferation of feral horses, as well as the overgrazing by ranchers paying a pittance each year for the privilege of producing equally damaging ecological ruin of natural habitats. The same topic, this time the damage done by brumbies (the Australian term for feral horses) in central

Australia, was taken up in *Outback Rambling*, a delightful book combining many astute observations on Australia and Australians interleaved with the feral horse problem. These direct field experiences and observations were conducted while helping his wife band finches for her behavioral research. The same opportunity occurred on a second visit to capture blackhearts, the local name for long-tailed finches, the ones with a black, heart-shaped bib. The book *Blackhearts*, published by Yale University Press, was also a double entendre, pointing to the setbacks the research encountered when three graduate students, ostensibly keenly motivated by field research, failed to come through under the occasionally discomforting conditions that fieldwork sometimes entails.

So despite the injustices he has suffered, we have here a highly motivated, productive, and imaginative geographer. Two-thirds of his professional articles have been accepted by major geographical journals, often despite their critical tone—something not worth mentioning in other fields, but less than usual in geography where the tradition of criticism has never been strong. And I think it is fair to say that despite highly varied topics—from prostitution, to feral horses, to Australian finches—Rich excels at two genres of writing. The first is criticism, at which he is merciless with his sense of accountability. But this type of professional writing, clearly colored by his experiences, may well be grounded in the firm conviction that it is the responsibility of the geographer to illuminate portions of our wonderful human and physical world by direct, astute, and so revealing firsthand experience, observation, and *talking* to the people involved.

Rich hasn't sat in an armchair to do his type of geography for years. As for talking to people, he tells an amusing story of a field trip in Costa Rica with Carl Sauer. Sitting by the roadside with students hanging on every word, Sauer explained that a field in the distance had been cleared within the past twenty-five years. Skeptical, Rich, and another student in the group who spoke better Spanish, wandered down to farmers working in the field. They learned that Sauer was wrong by a good fifty years.

Of course, Rich's view of research in human geography is not shared by everyone (why should it be?), but he makes a strong case by the sheer example of his own writing, a style of writing accessible to many interested readers far beyond the circle of professional geographers. And often his astute observations are recorded in his photography, for Rich is a fine, and some would say professional-level, photographer, with local and foreign exhibitions to his credit, again focusing upon people and their places, rather than the broader landscapes in which their all too human lives are embedded. Such field-based and people-oriented research marks one of his more recent ventures. With fluent Spanish, he eschewed the plush tourist hotels of Havana and "roughed it" all over the island, talking to and living with ordinary Cubans, finding (with rare exception) that as an American reaching out to them, he was made more than welcome by the ordinary people. We await a book on Cuba, perhaps unique for the directness and closeness of its observations, with great anticipation.

Conclusion? This book you are holding in your hands will be read with interest, pleasure, and puzzlement by many in the geographic profession and beyond. Assuredly, there will be notable establishment rumblings and expressions of outrage, as well as shrill voices from many post-fill-in-the-blanks whose relativistic "positionalities" have led them to being hoisted by their own petards. But I remind you that petards are small bombs designed to open up spaces previously inaccessible. May this book have the same effect.

University Park, Pennsylvania
2000

Acknowledgments

JOHN Short went to bat not once but twice for this book with Syracuse University Press, and for this effort we are most grateful, for from the beginning we've felt that the odds of any university press publishing this book were no better than one in fifty. We are fully aware of what we have written in what follows and how it will be perceived and judged by most.

Irene Vilar-Cuperman at Syracuse University Press also deserves a hearty thanks for backing the book and being willing to take a chance, not easy when you're relatively new to the job.

Annette Wenda did a splendid job of copy editing. We feared that any copy editor would love nothing better than to change our biting, ironical, or mocking tone, and demand that we forget that we had ever been influenced by Lenny Bruce and had grown up in the street. But to our good fortune, she let us be ourselves. For this we are most grateful. She did, however, catch inconsistencies, grammatical slips, the missing footnotes, the occasional sentence that didn't quite make sense, and redundancies not deemed necessary.

Don Pietro fought vigorously to have this book published. Even while dying, he battled by telephone with Syracuse's senior acquisitions editor when, in round one, there was unanimous support for the book, including a second round of supporting letters, yet she said no. The woman was, to judge by her correspondence, just plain scared of the critical, offbeat nature of the book.

Don Pietro was invariably quick to respond to these essays when they were first written, even before there was any thought of putting many of them together as a book. He gave us detailed written criticisms, and he then went over most of these essays a second time when they were in book form. Now we regret only that he is no longer alive, for we very much miss this trusted ally and friend, whom we never met in person.

Quetzal, another trusted friend, also fought hard to get this book published. And, like Don Pietro, and following him, he went through the book line by line. He made the valuable suggestion that we add in more about our histories, material to be found in what is now the final chapter. Much is made obvious that may be anything but obvious to people who know little or nothing about us and some of the issues we raise concerning censorship and criticism of the power elite.

Our friendship with Jacquot began more than thirty years ago. He has often taunted us to read this or that article or book, knowing that he could then expect some kind of pointed response. Had it not been for one particular provocation, regarding Ed Soja's book *Thirdspace* (see the chapter "Academic Brahmins"), it's quite possible we might not have written the two dozen or so essays that followed in fairly quick succession (not all of them included in this book). Jacquot also did battle with people at Syracuse on our behalf, and he knows that we're thankful indeed. Appropriately enough, it was Jacquot who suggested the title of this book.

John of Oz is, we daresay, our biggest fan and supporter for the kinds of critical essays that follow. He detests the pretensions, bogus claims, and cowardice of the academy every bit as much as we do. Like Don Pietro, he was always quick to respond with comments on these essays. And like Quetzal, he too pushed us to include more in a conclusion about our histories, arguing that to expect people to "get it" without giving more explanation is to assume more than can be expected.

There was another person to whom we once looked for feedback, and who deserves acknowledgment. He too gave us reactions on these and other essays when first written, sometimes with long and fascinating bibliographic asides. And—put bluntly—he loved the "hits," and he thought that the people we were criticizing deserved exactly what we were giving them. But all of this encouragement, and a friendship, came to an abrupt end when we wrote one of these essays on his work (included in the book). Apparently, he didn't believe us when we told him that when it comes to matters intellectual, everyone is fair game. When we wrote the essay about Pablo, we sent it to him and invited him to give us as much feedback as he desired, noting that anything he sent that was reasoned and reasonable we would take note of and integrate into what we'd written (a courtesy implicitly if not explicitly extended to just about everyone written about in this book upon sending by e-mail what we'd written). But Pablo could not get beyond petulant anger, heaving buckets of personal bile at us in a string of rapid-fire e-mails. Even then, repeat requests by us for criticism directed at what we had written, not how he felt about us or how he wanted to tell us how others felt about us, fell on deaf ears. With time, we saw that Pablo is an archetype in geography and beyond, if not in responding, then certainly in attitude. It's all, foremost, about "friends" and friendship, which means that friends of any stripe are off-limits when it comes to hard-nosed intellectual debate. But where, one must ask, does friend end and everyone else begin in the intellectual arena, particularly in a discipline as incestuous and insecure as academic geography?

Oddly enough, the best and most thorough criticism I received on this book came after it was accepted for publication and before it went to copyediting. Off and on for a couple of months, I had a vigorous e-mail exchange with a person who insisted on remaining anonymous and not being named in the acknowledgment. This is unfortunate, because this person made the book much stronger, and even more coherent; and I believe this person may have as good a sense as anyone of how this book will be received by geographers. I did not agree with everything suggested, but in the end I incorporated, in both major and minor ways, well more than half of the criticisms offered. Kudos to you, friend.

Fortaleza, Brazil
2002

GEOGRAPHY *inside out*

Introduction

The Critic and Criticism

IT is now just about twenty years since John Agnew and I wrote a small book titled *Order and Skepticism: Human Geography and the Dialectic of Science.*[1] A principal point of the book was that any science or academic discipline that's not self-critical and constantly vigilant about what it's producing is in danger of putting forth and perpetuating lots of self-fulfilling propositions, and more than a small amount of nonsense. This point is one that twenty years ago, as now, strikes me as so obvious that at times I wonder how John and I managed to make so much of it, or why it had to be said at all. But, alas, it not only had to be said then, but still needs to be said, and emphatically so—and, I fear, again and again ad nauseam.

Even by the second year of graduate school, I'd become aware that I could not expect much in the way of criticism, or responses of any sort, to whatever I wrote. If I gave a copy of a paper to eight or ten people that I knew and whose opinions I thought might improve my effort, I would be lucky to receive one response, even in those cases where I called the people friends.

By the time I was a year or two out of graduate school and teaching in a geography department, at a time when I had not yet cemented my reputation as someone certain to respond to almost anything, and perhaps with a hot missive, nothing had changed. I'd write a paper and ask several of my colleagues in my department, and people I knew in other departments, for critical comments. I'd invariably stress the word *critical*. As the weeks went by, I would find little or nothing in my mailbox, no colleagues knocking on my door to discuss my paper, no telephone calls to share an idea, a thought, a suggestion. Each time after one of these experiences, I'd tell myself not to bother in the future with sending around what I wrote—but I always did.

1

After my articles would appear in print, and often in journals with thousands of subscribers, I would get a few requests for reprints, and a couple of compliments. That was it. With the exception of an occasional letter sent to a journal editor, to which I would usually respond, no one would take me to task, ask me to elaborate, or suggest how I might've improved what I'd written. I often felt as though I would have been just as well off to write whatever was on my mind and then file it away in one of my file cabinets (a practice that I actually began as early as the mid-eighties).

The books I've written have had a similar fate to my journal articles. Except for book reviews, to which I've never been allowed by editors to respond, I haven't received enough reactions of any sort on all my books combined to provide even a good half hour of reading.

There is one area where I have consistently gotten feedback on what I've written, and that's from journal editors. But rarely have I gotten more than the message that they are conduits for what they have received from reviewers, and that they are conservative gatekeepers—this theme most of all. With a rare exception, academic editors have reminded me that, first and foremost, they represent the middling disciplinary norm, a norm in which language, subject matter, and treatment are, if nothing, huddled well inside one standard deviation. Thus, anything said by reviewers, no matter how off the wall or wrongheaded, is fodder for an editor when deciding whether to publish a piece. Logic, reason, and documentation, no matter how much I have provided by way of lengthy and repeated responses (and I have heard many others speak similarly), consistently fall on deaf ears.

Of the books I've written, I cannot recall that a single major idea was challenged in any one of them by either a reviewer or an editor. In my book on prostitution, in the one on wild horses, and in the collection of essays on outback Australia, I received suggestions to move or seriously rewrite no more than a grand total of half a dozen paragraphs in some three hundred thousand words. In my most recent book on fieldwork and bird ecology in outback Australia, I was "forced" to eliminate the sum total of one run of dialogue and one section running to about fifteen hundred words.[2]

What, briefly, do I make of all this?

It would be easy to assert that for all that I've written, I've simply had little or nothing to say, and therefore anyone responding, either when I was a graduate student, years later, or more recently, was, in effect, saying, Your articles, essays, and books are so consistently empty of content and reason that they're not worth bothering with. This hypothesis might be reasonable but for the fact that: (1) I've written about a lot of different things, and several pieces have appeared in high-profile places; and (2) a fair bit of what I've written has been "hot" or "controversial," and therefore ought to have invited reactions if only because I have gone against the conventional wisdom, in more than one instance angering geographers in the same way that spouses fight with one another and wind up shouting, throwing blows, or going to divorce court. If others I've known have had a different response over the years, then all this lack of interest in critiquing what others write might be all about me. But from all that I know and have heard, what has happened to me is commonplace, nay, extremely widespread. And if someone wants to claim otherwise, then I'd need to see good and convincing evidence.

With regard to my particular case, it is certainly true that I have picked up a reputation for answering at length and not always with a lot of love and tenderness. This reaction may have had various consequences. People who know about me feel they have enough to cope with without wanting to deal with hard criticisms, fearing perhaps that I might retaliate by doing to them what they've seen me do to others. Or they may be insecure about their own ideas and don't want me, or anyone else, to respond to what they say, least of all to show just how shoddy their reasoning is or lousy their ideas are.

I certainly don't believe yet another possibility, namely, that I'm "right" in every instance and therefore there's really nothing to say by way of criticism about what I write, more often than not with the speed of a journalist facing a do-or-die deadline. I may well be above average in intelligence in a sample of academic geographers, but I'm fairly certain that whatever I believe about the generally poor quality of articles and books written by geographers that I read, there are some quite bright people around who are producing some very good work.

When academics are graduate students, they are learning and not often sure of their footing, and in any event don't want to offend anyone who might be a future ally or entry point into a job. In other words, besides whatever insecurities may precede and accompany the learning process, they're simply trying to look as good as possible, which very often means being cautious about who or what is criticized. The calculation is a social one, not an intellectual one. Little changes with the first job, and in fact little changes all the way to tenure. Scholarship, honesty, truth, integrity, an interest in the limits of one's own mind: all are subordinate to tribal norms and concerns, and real and imaginary expectations about being liked, doing what has to be done to get a fair share of tribal merit badges.

By the time academics are tenured, their patterns of behavior are well established. If by this time they've not been in the habit of critiquing the ideas of others openly and taking chances with what they do and don't know, and sometimes discovering that they're wrong, then they're not suddenly going to change just because they now have lifetime job security. Inertia and habit are, and even somewhat independent of social or tribal norms, determining.

Then, too, once tenure's in the bag, there are now all those other things that get in the way of spending time critiquing what others are doing. There's the demanding wife or husband and kids; the chores of shopping, cleaning the toilet, and getting the car repaired; and all those committee meetings to attend and lectures to repeat. There may even be the desire or need (for money as much as elusive fame) to write an article or two in the next year, perhaps get started on a book. In short, it's now way too much trouble to think hard, and write what you've thought about someone else's thoughts. In fact, it doesn't take much thought at all to conclude that it pays much bigger social and personal dividends to simply not respond at all to a request for a critical review, or respond with a comment of this sort: The paper's great—I love it! After all, who doesn't love flattery and flatterers, whether in bed, the boardroom, or the halls of academe? Put differently, everybody's happy because, after all, being an academic is not, all protests and hyperventilating to the contrary, fundamentally about refining ideas, getting closer to truth, enjoying a good intellectual tiff, identifying beauty in the well-constructed argument, being honest and showing integrity, or pointing out the fakes and frauds that ought to be freighted out of town. No, it's all fundamentally about getting along and collecting friends, and acting tough and critical only when there's an upstart graduate student to assert dominance over or a well-known consensus that so-and-so is an undesirable lout and his ideas are unquestionably inferior (reasons at this point need never be given and rarely are).

One consequence of all this social worrying, endless temporizing, and, yes, intellectual cowardice—and I have sketched only some of the reasons for the sad state of affairs—is that academic geographers (and they're not alone in the academy in this regard) get precious little practice in critiquing the works of others. No different from writing, playing the piano, or learning how to do good photography, you learn by doing. You do not learn how to critique by reading or watching and then creatively imagining that you could have done as good a job, or better. To believe so is akin to reading a good novel and embracing the flightless fancy that you could do as well by merely putting pen to paper.

A lack of practice at critiquing the works of others is almost certain to make itself evident when called upon by editors to provide evaluations of papers and book manuscripts. Lack of practice is almost certain to produce second-rate criticism. It cannot consistently be otherwise, and to think otherwise is just another conceit on that very long list that defines a great deal that's wrong with the academy.

Some of these difficulties could be corrected if at least two conditions held: (1) editors were unusually bright and perceptive across a range of methods and subject matters, and honest people of principle; and (2) editors took seriously, or much more seriously than they generally do, the responses to reviews received by authors. But very few editors that I have encountered in the past thirty years in academic geography have been all that savvy, honest, or willing to listen hard to what authors have to say about all the drivel they receive that parades under the name of "reviews."

The problem is further compounded because editors, in my experience, often think they are something special, brighter, more perceptive, and more wide-ranging than anyone else around. In fact, rare is the editor who's not full of himself, and living a life of self-delusion about his talents and the everlasting significance of whatever appears in his journal. This arrogance and shortsightedness carry over into the license given to reviewers, with rare exception a license that allows them to say anything in any way they so desire, and get away with it, and to even charge all over the place with accusations and assertions without providing more than a scintilla of evidence for their claims. In this kind of climate, defined, aided, and abetted by agenda-driven editors, authors are all but defenseless. It matters not a whit that the author may know considerably more about a subject than the editor or any of the reviewers. When, to all this, the baneful and cowardly practice of anonymous reviews is added, as well as the practice of editors making up or faking reviews to fit their own ends—practices a lot more common than anyone cares to admit or imagine—the odds against an author not in the mainstream or with just the right connections (meaning a stable of people who are equally compromised) getting a fair hearing are poor indeed.[3]

If it further be the case that an author is not part of one of the many subgroups that define any discipline that he is taking issue with, then, by the very fact of being an outsider, he will be ignored. In other words, being on the inside means not taking issue with the "paradigm"—keeping quiet, keeping your nose clean, keeping dissent to a minimum—whereas being on the outside means not *knowing* the paradigm, so say the ones on the inside, and therefore, by definition, being incapable of writing a legitimate critique or having anything worthwhile to say. Members of disciplinary subgroups are like alcoholics who are interested in hearing only what other alcoholics have to say about drinking.

In my particular case, not only am I an outsider to just about anything I take issue with in geography, but I also do not hold a university position within academic geography (notwithstanding a Ph.D. in geography, I am a tenured faculty member in a department of ecology and evolutionary biology). In a sense, I am doubly on the outside. It is easy not only to ignore what I might say by way of criticism but also to claim that I am, well, just a bit mad.

When the problem is further compounded, as in my case, by a reputation that precedes me and makes me larger (or smaller) than anything I am or have ever done, then this just makes it all that much easier to see me as a "crank," a "nut," bitter or angry about this or that thing in more ways than anyone cares to take the time to specify.

Me and my case aside, the result of a discipline such as geography that lacks a critical tradition—and it does indeed—is that it is a discipline profoundly impoverished, and destined to remain so. It is a discipline that, on the whole, is an intellectual desert. It is a discipline littered with intellectual crap: articles and books that are poorly researched, poorly reasoned, poorly documented, poorly written, and trivial or pointless—which is by no means the same as saying there are not a number of good articles and books to brag about. The bad stuff remains in place precisely because there is no ongoing and vigorous critical tradition.

Very few people can long tolerate someone who's complaining all the time. If one's wife or girlfriend (husband or boyfriend) is always carping about matters large and small, most of us—and that includes me—are going to tell her to cool it or shut up, and if she doesn't, then she's going to be looking for another spouse or someone else to sleep with. But here we're not talking about a person's intolerance for bitching, negativism, call it what you will. At issue is quality of intellectual output, a serious matter in my mind.

When I get down to the business of examining what purports to be serious intellectual output in the form of an article or book or whatever, I dramatically shift gears. Then, with rare exception, I don't have the least bit of interest in whether I offend someone by what I say about what he has written. I am interested in calling a queen a queen if she looks, behaves, and talks like a queen, and calling someone a lousy scholar and intellectual impostor if this is what all the evidence before me shows. Those academics who don't want to play the intellectual game as I and a few others I know do should do something else: play Ping-Pong, play with the kids, play with themselves.

I have heard the argument that the house of geography is plenty big enough for all kinds of studies, points of view, and what-have-you, and that by being critical (which most people translate to mean negative) I am not only parochial but insufferable as well. This kind of fundamental misunderstanding of the role of the critic and of criticism is just another way of simply making apologies for mediocrity, of elevating a social and tribal norm, and perhaps considerable personal shortcomings, above an intellectual norm. It is an attitude that shows a lack of discrimination; it says, in effect, I don't care about making distinctions between a Picasso original and a beach-fair cardboard imitation. It's a way of saying that we ought not to pay attention to the vast difference between romance novels and, say, the work of Borges or Coetzee. I've nothing against romance novels; I just don't care to confuse them with the best that serious minds can produce.

As long as professional academic geographers take lightly their duty of putting on intellectual gloves and getting in the ring to see who can go ten or fifteen rounds, and are willing to suffer knockdowns and the occasional knockout, academic geography will not only remain the intellectual desert it is, but also invite further erosion of its long, precarious existence.

I expect that many of these essays will be received with a fair bit of hostility. Some will say that my approach is offensive, crude, unacceptable, and unprofessional, and therefore I am not to be taken seriously. Thus, to these critics it's all about form and appearance, not substance. This kind of reasoning, by "reasonable" and "professional" people, is, to my way of thinking, not to be taken seriously. It's all akin to oppressed women, exploited minorities, and so-called terrorists (Palestinians come quickly to mind) being cavalierly dismissed because their methods were and are considered offensive, crude, and unreasonable—by all those "reasonable" people in power. A good rule of thumb, well learned by the sixties generation, is: distrust on principle anyone who claims to be "reasonable."

Some will say that I have picked on easy targets, and that there are all kinds of good geography that one can point to. In fact, virtually all the geographers whose work I critique in this book are high profile: they are full professors in major universities, they are people who have substantial publication records, and they have had, and are still having, a genuine impact on graduate students and on what other practicing academic geographers have done and are doing.

To be sure, there is some quite good and solid work in geography, and even some that is excellent, and there is no shortage of geographers who have been eager to extol these efforts. But what is so sorely missing—to repeat at the risk of being redundant—is that identifiable clearinghouse where the few good pieces are plainly identified and separated from all the dross. What is so striking to me is how much bad, irrelevant, and pretentious geography is being produced by a discipline that now, to judge by its major organization, the Association of American Geographers, has some seven thousand members.

A final telling note is in order, a microcosm that I think nicely illustrates the sad state of affairs in which geography finds itself in the year 2001.

After this book was accepted for publication and was awaiting assignment to be copyedited, I learned that Robert Sack was quite unhappy with what I had written about him in one essay in this book ("When Logic and Language Go on a Walkabout"). He apparently was also unhappy with what I had written about other geographers, essays that he'd either seen after I'd shared them with others in a draft form or heard about through the always reliable, gossipy grapevine. As luck would have it, he found himself in a position to make these views known to someone who had been involved with the book's review and approval for publication.

This individual pointed out to Sack that any objections he and others might have could be expressed once the book was in print. Sack was furious, and when it became clear that the individual showed no inclination to revise his assessment of this book or influence its publication, Sack informed him that he wanted nothing more to do with him ever again.

Sack, it seems, reasoned that the individual would see that it is widely known in geography that Richard Symanski is an outcast and something of an outlaw, and that nothing would be gained by staying with a decision that seemed to favor Symanski's interest. Robert Sack not only is a thirty-year-plus veteran of the highly respected University of Wisconsin at Madison Department of Geography, but also holds two named chairs, Clarence J. Glacken and Bascom Professor, which is highly unusual. To boot, he is a professor of integrated liberal studies. The university considered Sack highly enough to acclaim him as someone who could speak authoritatively for, well, liberal studies: the whole of the social sciences and the whole of the humanities, and great stretches of history too. In short, the individual had to know that he wasn't hearing the petty grievances of a journeyman academic.

Shortly after receiving this information, I (Korski) wrote a letter to Sack. I let him know that I considered his behavior an infringement of my First Amendment rights, an abridgment that I found particularly galling because Sack and several other members of the University of Wisconsin at Madison's geography department had written a protest letter to the AAG Council in 1982 when it censured my book *The Immoral Landscape: Female Prostitution in Western Societies* (see the chapter "An Anatomy of Academic Censorship"). Sack obviously understood—at least he did twenty years ago and when involving someone else—the importance of the First Amendment. Or did he believe that the First Amendment issue was important only when it did not involve him personally?

In the letter, I offered Sack what I thought was a good deal. If he would formally apologize to me and to the individual concerned for his misguided behavior, then I would forget the whole matter. I offered him more. I sent him the final version of the essay I'd written that appears in this book, and I invited him to rebut what I had written point by point. I said that any valid points he made would be incorporated into a revision, and if he made enough good points, or hit on something absolutely crucial that I'd gotten wrong, then I would take the essay out of the book. I went further. I sent, via e-mail, the identical letter to most of Sack's colleagues, and with each one I included the essay I'd written about Sack. I invited one and all on the Wisconsin faculty to come to Sack's defense and write a rebuttal.

I told Sack there would be a price if he refused my offer, and that price would be the very narrative you, the reader, are now reading. I ended the letter with the following words: "The ball is in your hands. Let's hope, against 100 to 1 odds, that you and those around you know what to do with it." I heard not a word from Sack or any of his colleagues concerning this issue.

This question may arise: Did Robert Sack in fact violate the spirit and intent of the First Amendment? He certainly had the right to tell anyone he knew that the essay I'd written about him was wrongheaded, misguided, or whatever. He also had the right to express similar sentiments about other essays I've written. But he did not have the right to throw a veiled threat at someone intimately concerned with the book's review and final approval, saying that he wanted nothing more to do with him ever again because he would obviously do nothing to prevent the book's publication. At this point, Sack found himself standing squarely on that turf called censorship. And the reason this position is where he found himself, rather than on some innocuous ground where all of us vent anger to equals and to ourselves when alone, is precisely because of the social and political context of his statement: to reiterate, the fact that Sack had been at the University of Wisconsin at Madison for thirty-some years, the fact that he held two distinguished chairs, the fact that the geography department had long been nationally recognized and held in high esteem within the university, and—this too—the fact that the most recent chancellor of the University of Wisconsin, David Ward, was not only a geographer but also a friend of Robert Sack. The individual that received the veiled threats was fully aware of these facts and what they would lead any reasonable person to conclude.

Alas, when the person did not budge from his position, Sack went much further. He made this person's job difficult, with consequences that had nothing to do with this book. He behaved in a way that the individual affected described as being "extremely vindictive."

Robert Sack had ample opportunity to try to convince me that I am badly mistaken in my assessment of his intellectual abilities. Rather than work within a critical tradition and avail himself of my offer to engage in a dialogue that I initiated, he instead chose to use his social and political power to try to suppress what I have written. This reaction is the mark of someone who lives not by reason but by prejudice. And prejudice of this sort is everything that universities do not, in principle, stand for.

Recently, I heard there was a need for *meseros* (waiters) in the Adelita Bar in Tijuana. It has occurred to me that the Adelita would be a great place for Sack and his ilk to pimp their unparalleled intellectual wares, their ideas about intellectual exchange and criticism, and their retrograde attitudes about freedom of expression.

Part One ⊚ Postmodern Truths

> The author's career is a course whose record is his work and whose goal is the integral text that adequately represents the efforts expended on its behalf. A text is the source and the aim of a man's desire to be an author, it is the form of his attempts, it contains the elements of his coherence, and in a whole range of complex and differing ways it incarnates the pressures upon the writer of his psychology, his time, his society.
>
> —E. S. Said, *Beginnings: Intention and Method*

First Holy Communion

Korski and Symanski

A certain pleasure is derived from imagining oneself as individual, of inventing a final rarest fiction: the fictive identity. This fiction is no longer the illusion of a unity; on the contrary, it is the theater of society in which we stage our plural; our pleasure is individual—but not personal.

—Roland Barthes, *The Pleasure of the Text*

I met him in a whorehouse. It was at the Villa Joy at the end of Baud Street in Winnemucca. It all began when two buckaroos came around. I knew they were trouble. I'd known them from my days working on the Spanish Ranch, north of Elko. They demanded drinks and they demanded girls. But the working girls were with men in the rooms, or they were off for the night. The buckaroos got louder in their demands, and Penny, who was the madam, told them to back off. Finally she told them to leave. They wouldn't.

He turned on his bar stool and said, I think you should leave if Penny asked you to.

One of them came over to him and told him to step outside. He turned around to face the working girl he was with, and the buckaroo slugged him on the back of the neck.

He picked up Penny's birthday present from the girls. It was lying on the bar. The present was a nine-inch piece of hardwood covered with barbed wire. It looked like a penis. It was meant to look just like a penis.

He swung around and hit the buckaroo on the side of the neck with the barbed wire penis. Blood squirted from his neck. The buckaroo yelled and swung and missed, and then he got kicked in the balls. He went down.

The other buckaroo charged.

He punched him in the face and kicked him in the groin. He then picked up the buckaroos one by one and threw them out the front door.

Penny called the sheriff to come and get the troublemakers. Two came. They wanted to know what had happened. Penny said the bloodied bucka-roos had been loud and drunk and caused trouble.

The sheriff and his deputy handcuffed the troublemakers and put them in their car. They then checked the girls' work cards. They talked to all the girls, looking for other stories. They got none. They looked in the empty rooms. They didn't check the rooms where girls were busy with customers.

He and Penny were good friends. He came often that summer, she said. He taught her some Spanish, and they told each other bawdy jokes. She didn't know who he was or what he did. He didn't say much more about himself than the girls said about themselves. Penny liked him for his po-liteness and the way he behaved around her working girls.

I was just like him in many ways. I liked Penny and the girls at Penny's the best on the Line. They were young, gorgeous, and playful.

We talked and exchanged stories about wild horses, and later he rode with me and showed me some things I didn't know. He showed me how to capture wild horses. We rode and camped in the Clan Alpine Mountains, the Granite Range. We rode often in the Limbo Range where I'd worked and he worked.

Many years after our first meeting at Penny's, I had an on-again, off-again affair with his wife. His wife said she didn't know when he'd become what he was. He had been distrustful of authority since she met him. He never wanted to join much of anything. She said he was easy to get along with until somebody gave him shit. Then he was difficult. He didn't have much of a sense of restraint. He said all the time that life was a short stick.

There were a lot of stories about him. There was a professor at one uni-versity whom he allegedly beat up in an alley. This guy called himself a queer and was proud of it. There was a professor at another university whom he allegedly slapped around in a student café. His wife said all these stories about him beating up people were false. She said other stories about him that she heard or he heard were false too. People come to the kinds of truth they want to believe and call it everybody's truth, she said he said he believed.

I once talked to one of his brothers. He told me that he had changed to what he became when he was seventeen. That was the year he got kicked in his right eye and more or less lost it.

I once talked to his mother. She said he and his father were at each other all the time. His father had all the answers and was always right. The son hated this attitude. He hated any kind of authority or exercise of brute power.

Once I dug through piles of manuscripts and notes that were his. I found this one. It was undated but titled: "What His Father Did." I presumed it was about him.

When he was very young and he asked his father why he had to do something, he'd invariably say, Because I said so. I run this house and I know best, so do it.

No matter how much he persisted, his father never gave explanations. As the son got older and thought that he had more right to know why he had to do things that struck him as wrong or illogical, he became more insistent that he be treated as an equal. His father did not change, and indeed he continued to punish his son as he always had when he did not keep quiet or follow his dictatorial demands to the letter. The punishment took one of two forms. Either the son was locked in his room for the day or, more commonly, his father would force him to kneel on the hardwood floor between the living room and the dining room with his legs stiff and his arms straight up in the air, holding large apples or oranges. Sometimes his father would leave the house and be gone for hours. He would ask his mother to free him from his predicament. But she could do nothing. She was beholden to her husband's every whim.

Many years later, as the son lay beside a French Avianca stewardess in her elegant third-story Bogotá apartment, she asked him what his parents had meant to him.

He said, I think I would do almost anything for them.

She found it hard to believe that anyone could care this much for their parents. She said, They never treated you badly?

He told her about how his father had punished him.

When did he stop treating you so cruelly? she asked.

There came a day when he was no longer bigger and stronger than me, he said.

But you still accepted your father's authority?

Never again. To this day, I give him no more than I give to anyone. Heart in matters of the heart, reason in everything else.

You were scarred by the experience, the way your father mistreated you?

He nodded.

She asked for an explanation.

He said only, I accept no authority on principle.

None?

None. He then told her about the Dominican nuns he'd had in Catholic school and that he hated their authoritarian ways.

Presently, she asked him to tape emeralds to her thighs and the skin around her anus. Within the hour she would be leaving for Miami. While he did as she'd asked, he told her that his father had little education, that his values were Old World, that he'd been repeatedly beaten by an older brother as a child. He said that in his own peculiar way, his father loved his mother and had fed and clothed him and his two brothers as best he could. He said, He simply does not live in a world of reason, the kind to which I aspire.

It will be different in a university? she asked.

Of course, he said.

So you don't believe you'll find anyone like your father in American universities?

A rare one here and there, I suppose. But they'll be easy to avoid. Reason and principle will rule.

Then, he was naïve.

All his legal documents are in the name of Richard Symanski. But his wife of many years more often than not calls him K or Korski, and she often addresses loving cards to him as Korski. Around the house, she calls him by these names with affection—the strongest kind he often thinks—and with a certain kind of unspoken understanding about how she feels when using these names and how she sees him.

For longer than he can now remember, all dinner reservations for Richard Symanski have been made under the name Korski. When he travels, and especially when he finds himself in places where he might find trouble, or is aware of a mood shift that makes him more aggressive, he is Korski.

He has a handful of friends scattered in this country and abroad with whom he corresponds on a fairly regular basis—some by e-mail, some by snail mail. For most of them, what they receive from him is usually signed Korski.

Whether Korski is someone different from Richard Symanski in their minds is hard to say. Most of these friends make some kind of distinction, for sometimes they address him as Rich, other times as Korski.

Horse Thief

It's Borges, the other one, that things happen to. Years ago I tried to
free myself from him, and I moved on. . . . So my life is a point-
counterpoint, a kind of fugue, and a falling away—and everything
winds up being lost to me, and everything falls into oblivion, or into
the hands of the other man.

   I am not sure which of us it is that's writing this page.

—Jorge Luis Borges, "Borges and I," in *Jorge Luis Borges:*
   Collected Fictions

HE stuffed the last handful of crickets into the pouch on his belt, put the
pen flashlight in his pocket, and hiked down the pine-clad hill toward
camp. As the small fire came into view, a flaring candle against the brittle,
blue dawn, he paused to tune in the cacophony of chirps of several hundred male crickets, now warmed by the approaching morning and the
closed space of their prison. He registered their noisemaking, and he cal-
culated that the temperature had risen roughly three degrees since he had
gotten out of his swag in the black of night to go after them. He shook the
bag of ant eggs in his left hand and dropped his other hand onto the long
pouch of crickets. He thought of how much he relished these mornings:
the awakening air, the unpopulated vastness, a land he could almost call
his own.

   He brought to mind a photograph he'd once seen in the Smithsonian. It
was of Quanah Parker, the famous Comanche, master of the plains and incomparable buffalo hunter, the Serpent Eagle who wore tightly bound
braids all his life and was buried alongside his mother in an unmarked grave
near Cache, Oklahoma, dressed only in a tailored European jacket and a
derby hat with a large hole in its brim, his braids gray, taut, and full of lice.

For a long time, he sat in the dark listening to their stridulating sounds, the high-pitched trills and buzzes, the courting and mating calls. The paint in Dry Creek Canyon came to mind. The night before, he had saddled up Poison and ridden him up through Sheep Canyon and around to the western approach to Iron Mountain, searching for an even better mustang than the one he'd lassoed and hobbled. After setting up camp, he had set off in the dying light to find the cryptic ant mounds. And then, not long after dark, he came upon the cricket leks. The leks were large and rich, as big as any he knew in this part of the Cripple Creek Mountains. When he finished examining the leks, his curiosity temporarily sated, he knew that as long as he harvested them with care, and as long as men and their domesticated animals did not trample these remote grounds, he could return year in and year out for as long as he lived to satisfy a taste shared by fewer than one in 50 million Americans.

He put some cooking oil into the heavy black frying pan, got out the pancake mix, grabbed a bag of coffee, and poured an ample amount into a gallon coffee tin. He added water to the coffee, stirred the batter until it became gluey. He threw in the ant eggs and two handfuls of crickets, stirred some more. Soon the batter was ready. Feeling hungry, he licked his upper lip, whetting the long whiskers that he hadn't trimmed since leaving home nearly three weeks earlier. He poured himself a cup of dense black coffee, and as he savored the thick brew Fifi came to mind.

He broke camp and got ready for the slow, easy ride into Dry Creek Canyon. Now he remembered how long it had taken him the day before to cut the adorable paint away from her band without giving hard chase. It was early spring, the one time of year that made him think twice about chasing mustangs. Foaling was heaviest at this time of year in north-central Nevada, and a stressful chase could result in a miscarriage. The few times that he'd caught mustangs in the spring, he'd done so at water traps where there was little danger of a mare miscarrying. Now, however, he didn't believe that he had much choice. Daryl's birthday was only two weeks away.

Even before he'd heard the bad news about the cancer, he'd decided to give Daryl a mustang for his birthday. Then, last summer when he took Daryl to the rodeo in Elko and the young boy fell in love with a handsome paint, he set his mind to roping one of the few mustang paints with good conformation that he'd seen in these mountains. On two different occasions in the past month he'd tried to catch her at a water trap, but each time she somehow managed to elude him. Once, just after she was about to follow her mother into the circle of junipers and barbed wire that enclosed a well head and water trough on the Flowing Spring Ranch, the band she was with got spooked by the eerie cry of a mountain lion. The other time, the paint had followed the example of the harem stallion and fled. They would drink later, when there was nothing strange in the air, nothing to be wary about.

Korski spurred Poison, they shot up the rise, and shortly they were sitting on the lip of the draw that led into Dry Creek Canyon. Korski lifted his head and sighed. He stared at a wrinkled contrail to the west—another one of those damn booming jets from the naval air station. The glistening sun made him wonder how many Northern Hemisphere sunbathers would have their retinas permanently scarred in the next year.

Poison whinnied and wanted more attention, so Korski gave it to him by scratching his mane and whispering in his ear lines he'd once read in one of Thoreau's journals: How vast and profound is the influence of the subtle powers of Heaven and Earth! We seek to perceive them, and we do not see them; we seek to hear them, and we do not hear them; identified with the substance of things, they cannot be separated from them.

Poison's ears rattled like loose windows in a hailstorm. He understood exactly as much as moderately inbred horses are capable of understanding: the loving attentions of their attentive masters, and not a whole lot more.

Korski pushed up his worn wide-brimmed hat and quickly ran his eyes around the canyon. He took note of abandoned gold mines, including one wherein lived a seven-foot rattler whose mating habits he was following with consuming interest. He scanned the ridges, pausing here and there to examine ledges and shadowy crevices. Nothing unusual caught his eye. Looking down the canyon, north of tiny Pin Cushion Butte, he put his eyes on a bay stud and his harem. The stud was motionless, staring at Korski, waiting for the slightest indication that man and horse would break toward him. The mares and their young—two sorrel colts, two dark bay mares, and a gray yearling—were several yards behind the stallion. One of the colts had his head against his mother's stifle. The protective mare seemed scrutinizing, cautious. The other mustangs were farther back, off to one side, their necks extended to the ground, busily eating.

Slowly, Korski turned away from the band and, for the first time, gave his attention to the sidelined paint. Glistening in the strong light, she was no more than fifty yards from where he'd left her. She was lying down. She looked calm, and resigned. And what horse wouldn't be resigned when ropes and knots that had been tested since the time of the Phoenicians had robbed her of any hope of kicking the everlasting shit out of her human captor?

Poison pawed at the dirt and contracted her sphincter muscle just as a red-shafted flicker cried, Wick! Wick! Wick! The long-billed, gray-headed male flew out of a cringing juniper to Korski's right. Its salmon-red wings fully extended, its spotted belly proudly facing Korski, the flicker swooped low in front of Poison, then sailed down the canyon as if floating on a curving shaft of wind. Korski took note of his winged friend, then turned in his saddle to catch a glimpse of the female who had circled briefly above the tree and alighted on a knotted branch. He gazed at her for a long moment, looking for a familiar marking. Then he let loose with a sharp, loud, Klee-yer! Klee-yer! The bird shot from the branch and zoomed behind Poison and Korski. He smiled to himself and said to Poison, She's at it again, old boy. He laughed without restraint and slapped Poison on the neck. He laughed at himself, conscious of how often he talked to his horse when alone like now for days, sometimes weeks at a time. Poison shook his head, twitched his upper lip, flagged his tail like he knew exactly what was on Korski's mind.

Korski dismounted and reached for a pouch of tobacco in the right pocket of his sheepskin-lined denim jacket. He rolled himself a smoke, lit up, and dragged indulgently. He filled his lungs with smoke, and he began appraising the sidelined paint's conformation. She's a beauty, he thought. Rarely see one like her in these parts. Must have some of that good blood from those studs Panucchi was putting out.

Suddenly, Poison snorted, perked his ears, flattened them against his head. Korski's eyes darted among the junipers and piñons, around to the fissured black rock on the north side of the draw, onto the broken timbers leading into the abandoned mine shaft, back to the paint. The paint was now on her feet. She arched her neck and shook her head. She tried to break loose from the sideline. Korski moved several steps to his left to get a better view of the narrow opening where Dry Creek came into the canyon. He saw nothing. But something didn't feel right. He got down on his haunches. He looked for telltale signs among the trees, in the high sage in the low part of the creek bed, on the ridges. He listened for unusual sounds, missing sounds. He looked for the red-tailed hawk that often hunted the canyon for jackrabbits and rodents. The hawk was nowhere in sight.

Not satisfied that he was alone, Korski scratched the dirt at his feet and took some to his nose. The texture and the dryness of the earth, much drier now than the theater floor in Delos in midsummer, reminded him that the rains had not been good. Barely three inches so far, just under six the year before, four the year before that one. It had been six years since there had been as much as nine inches in this sheltered part of the Cripple Creek Mountains. Now Korski contemplated a nearby clump of Idaho fescue, then some Indian ricegrass. He shook his head. The canyon had only a fraction of the ricegrass that he'd seen the previous year.

He recalled the first time that he'd ridden into Dry Creek Canyon. That was the year ricegrass was everywhere; there was nary a spot where he could lay an open hand on bare earth. The following year, on the west side of the canyon, he'd seen the first cheatgrass. Within two more, the less nutritious weed of the Asian steppes had spread around to the northern and southern edges of the canyon. Worthless rabbitbrush, the water-loving and dangerous halogeton, and the deadly locoweed had also begun colonizing the canyon floor. And it was worse elsewhere in these mountains, the Granite Range, the Antelope Range, other ranges in north-central Nevada that Korski had come to know intimately.

His first time into Dry Creek Canyon he hadn't seen any mustangs. In the next two years, he had noticed that two stallions and their harems regularly came into the canyon for an early breakfast. Sometimes they stayed all day. Now there were times when he saw as many as forty or fifty horses chomping away on the ricegrass, the squirreltail, the needle-and-thread grass, other ice cream plants. In a mere five years, the antelope and mule deer herds that had long called the canyon home had diminished by almost half. Many had migrated farther north into the Fremont Range, where there were fewer cows and horses, more abundant food, more hunters, where they wouldn't have to wait for hours while mustangs tanked up. Korski pursed his lips, pulled on the long whiskers under his chin. The bastards are going to eat the canyon clean, he thought. Something's got to be done about these horses, the two-bit ranchers who listen to no one.

He took Poison over to a clump of junipers, tied him up, and gave him some water. He lay down in shade, read a few paragraphs from a novel written in Spanish that he'd bought on his last trip to Reno. He dozed, and long seconds became a long nap. After he woke, he got some water from a two-gallon jug that was tied to a strap on his tattered swag. He washed his heavily bearded face, brushed his teeth.

Feeling as civilized as he cared to, he put a wad of tobacco mixed with

finely ground green chili peppers in his mouth and walked Poison down to
the paint. She still hadn't completely shed her winter coat, and her long tail
and mane needed work. She was just plain ugly to the untrained eye. But
Korski liked what he saw: a strong back, haunches that were hefty and well
proportioned, eyes that were radiant, alive. Must go eleven hundred
pounds, he thought. Big for these parts. You'll make a great one for Daryl
once we get you broken.

He let the paint calm down. Then he roped her, dropped her, and took
off the sideline. He loosened the rope around her legs and slapped her hard
on the rump. She lamely struggled to her feet, shook her head, and whinnied. She took off on a full gallop. Within minutes, she was lost in the shadows of the confining exit along Dry Creek.

The paint was barely out of sight when Korski turned toward Raven's
Back and looked up at the bullish clump of piñons on the ridge top. He held
his gaze for several long moments before breaking into a supercilious smile.
His eyes high and fixed, he arched a wet wad of dark-green tobacco in the direction of Raven's Back. He opened his fly and peed into the wind. And he
thought: Fuck Gerbino.

Time of no concern, he sat on a rock surrounded by winterfat. He rolled
a fresh smoke. He looked back up at Raven's Back and tipped his hat. He
clucked to himself, then got up and checked a stirrup and got on Poison and
lazily rode out of the canyon.

In the corner office at the end of the long corridor on the fourth floor of the
federal building in Reno, Ernie Gerbino pounded the corner of Marv Pfeiffer's desk. He said, He knew, he knew I was there, that no-good son of a
bitch. I thought I had him this time.

Skunked again, huh, Ernie? And, to boot, you still don't know his
name. Pfeiffer chuckled and tipped back in his swivel chair. Marv Pfeiffer
was Nevada's chief law enforcement officer for the Bureau of Land Management (BLM), and Gerbino's boss.

That ain't the half of it, Gerbino said. I stayed up at the top of Dry Creek
the whole afternoon and all night and didn't get barely a wink. Damned if I
didn't have nothing in the truck to eat but a can of peaches. But I said to myself, No sir, soon as I leave, that bastard'll come and get his catch in the dark.
So I kept the scope on the horse all night long. Then he comes in the morning and talks to his horse and takes a nap, and I thought he'd never come and
get it. When he does finally come, he stands there at the mouth of the canyon
looking around and scratching himself. Like I got forever to collar him.

Patience, Ernie, patience. You get your check one way or the other.

So then he looks me right in the eye at three hundred yards, right square in the middle of the scope, like he's saying, Hey, cowboy, I didn't catch that pretty paint. I'll be damned if he didn't go down to her, take off the sideline, and stand there while it ran like hell out of the canyon. Then he turns back to me and spits a mouthful in my face.

He smelled you, Ernie. If you'd take a bath more often, you might have some luck.

Ten to one says—

—I'll give you forty to one you won't catch him.

Honest, Marv, I don't know what happened. I didn't breathe hard the whole time I was up there. Honest to Christ, I didn't move my eye from that scope.

The right one, right? Pfeiffer laughed and cupped his hands.

Gerbino pulled out his right glass eye and threw it at Pfeiffer. Catch, you old beer-bellied buzzard, he said.

The glass eye hit Pfeiffer's bulging stomach and landed in his hands. Gerbino turned the grotesque pink-and-purple socket toward his boss and stuck his finger in it. Tittering, Pfeiffer said, You ought to pop it out and try that someday when we're in there with the director and some of those Brooks Brothers honchos from Washington.

Yeah, sure, and get an early pension that won't feed my dog.

Hey, listen, Ernie. You do like I'm telling you, the director'll think it's candy and ask for more. Pfeiffer laughed, tilted his head back, and opened his mouth. He put the eye above his mouth and pretended to drop it in.

Give me my toy back, you old crock.

Pfeiffer sniffled a couple of times, then threw Gerbino his eye. You follow up that rustling problem we got in Smith Creek?

What you think I am, a magician? It took me three hours to get a lube job before heading out, and that don't count the time it took to hose the alkali off the engine. Then I had to—

—I know, I know, don't tell me. You had to go to the little boys' room, then pick up the horn and see what Mr. Snitch had to say. Right?

Willie hasn't let us down yet.

He won't, Ernie. Not as long as he sends you in one direction and goes in the other to do his own illegal mustanging. Ernie, you got your priorities all screwed up. You keep forgetting I have all these reports to prepare for the big man down the hall. Every time *Newsweek* or *Time* comes out with a piece saying someone shot a couple of mustangs up around Winnemucca or Battle

Mountain, the director gets on my ass. Wants to know what we're doing to enforce this Wild Horse Act. One conviction coming out of this office in the past six years doesn't exactly make us look like the Wyatt Earp gang.

So, our good friend Willie says, You want this law breaker whoever-he-is, he's on his way into Dry Creek Canyon. I seen him saddling up near the mill on the road into Fourfoot. Same as last time. He didn't have to tell me any more. I said, Balls, that rustling case that ain't going nowhere I can see can wait. I'm gonna cuff that son of a bitch if I have to lie up there on top of Raven's Head for a week.

I bet you would've too, Pfeiffer said. He opened a file cabinet drawer to his left, lifted his silver-toed boot onto the front lip, and thought of all the times he'd asked Gerbino why he wanted to catch a horse thief of no consequence.

Gerbino said, I want him real bad. But you know, strange thing is, I really like that son of a bitch. 'Cept for stealing our horses and doing God-knows-what with 'em, he's honest as the day's long.

You know that, huh?

That he's honest? Yeah, I think he's honest.

I didn't know they teach you to think them morality issues in those law enforcement courses.

Well, he don't own no ranch 'round here, far as I know. Whoever he is, he don't run around with a mouthful of bullshit saying he's got all his cows ear-tagged when he don't. Gerbino barely got the last word out of his mouth when he started coughing, then retching. He couldn't stop. His face turned red, and he bent over to catch his breath.

Marv Pfeiffer got up to pound him on the back. You go see a doctor like I told you?

Gerbino sat down and put the cigarette out. He took several deep breaths. He don't know diddly-shit about how good it tastes.

You're not going to know pretty soon, either, Pfeiffer said as he returned to his swivel chair.

Least I'll go out enjoying it. He laughed and started coughing again.

Pfeiffer took a pair of nail clippers from his desk drawer. He leaned over the wastebasket and began clipping his nails. Ernie, when you going to get this little old minor horse thief off your brain? How many times I have to tell you to forget him? He's the least of our problems. A few here, a few there, who gives a damn? Right now we got more friggin' horses in government corrals than we can ever hope to get rid of. Pretty soon you and me will be out there shooting 'em and taking the bulldozer to 'em.

What're they counting out there now at Palomino Valley?

Last I hear, better than a thousand. And there's another eight or nine hundred they're gonna round up in Lone Pine Valley soon. Now you tell me how the BLM's gonna get rid of that many jug-headed, slipper-toed horses. Dammit, Ernie, let's concentrate on a case we can get a little mileage out of. Something these high-rise judges'll take. Last time we took the prosecutors a three-horse case, they told us go play with ourselves. We want the right kind of publicity, we got to nail someone sending thirty or forty to the can in North Platte. Slicks, tails dragging the ground, no bill of sale, pictures, and witnesses. That sort of thing.

Gerbino said, You want evidence that'll sing in heaven, you get the BLM to send out a black-shoed senator to do the legwork. One little guy like me patrolling the whole state getting paid nothing can't do twat. Ain't me gonna kill myself for some greaseball bureaucrats. Two years ago, I worked my tail off on that Michelson case, and we got ding-a-ling in a bunny basket.

The Jackie Michelson case was a sore point with Ernie Gerbino. Michelson owned a small cattle ranch on the Black Rock Desert, north of the Pyramid Lake Indian Reservation. It was widely known that Michelson had been rounding up mustangs in northern Nevada since the late 1950s. He sold them to rendering-plant bidders who came in from California and Idaho. After Congress passed the Wild Horse Act in 1971, which declared that mustangs on public lands were a national heritage to be managed by the Bureau of Land Management, it became a felony to catch the horses. Michelson, like others long accustomed to mustanging, ignored the law. He caught and sold horses whenever the urge beckoned. Irked by Michelson's defiance of federal law, Gerbino spent nearly six weeks camped in the Calico Mountains, waiting to catch him with government horses. When Gerbino finally caught Michelson with thirty-six mustangs in a jerry-built trap, Michelson laughed, said that he'd be haying on his ranch the day the jury met. He got off with a small fine and a requirement that he spend a hundred hours repairing springs regularly used by mustangs. After the trial, Gerbino said to Pfeiffer, The law and the judges in this state ain't for diddly-shit. I'll chase the thieving bastards 'cause that's my job, but I'll go after who I want to from here on in.

Don't take it out on me, friend. I'm only saying it like it need to be said. Hey, you given any thought to going to Alaska to hunt grizzly? Marv Pfeiffer swung around in his chair, grabbed the 30.06 in the corner, and aimed it at a fat woman in a short print dress walking to her car in the parking lot below.

Not with the kind of pea shooter you got I'm not going.

I could take that two-legged trailer down there out at two hundred yards, Ernie.

Throw me your lighter, will you? I don't get another cancer stick in my mouth soon, I'm gonna fuckin' die.

He reached over to the corner of his desk, grabbed his lighter, and tossed in up in the air, toward Gerbino. Gerbino grabbed it, and Pfeiffer said, You ought to think about it, Ernie. We could take along the wives and leave 'em sewing in the cabin and then have ourselves a helluva time. I been thinking about an early pension, staking out a little land up there north of Anchorage to get away from all this paper and donkey work. That's what you ought to be thinking about. We go up there, you wouldn't have to get your blood pressure up every time someone takes a few oversized ponies. You have to admit, Ernie, that's some lousy goddamn law we have to work with. He rubbed the butt of the gun like he was searching his forearm for mysterious bumps. He swung around and took aim at a young, leggy woman in candy-red leotards. Bang! A good, clean heart shot, Ernie. Right between her two big titties.

On the way back, I was thinking of the perfect way to get him.

Jesus, what a shot! I oughta go see the director for a medal for that one. He fell back in his chair and put the butt of the gun in his groin. For*get* him, Ernie. Listen to what I'm saying. I know what I'm saying.

My ass, I'll forget him. I'll forget him after I got cuffs on him.

All right, all right, Ernie. Write up your report and make yourself look good. And, listen, give some serious thought to a grizzly hunt in Alaska, will you? My old lady's screaming every day to get her out of here and pay her more attention.

Gerbino slapped at Pfeiffer's words and reached for another cigarette in his left breast pocket. He cupped his mouth with his right hand, squeezed hard, and changed his mind. He got up and started coughing. He coughed all the way down the hall and into the bathroom.

Korski spotted the same paint, back with its family in the southern end of Ryan's Valley. He gave chase for the better part of three hours before he was finally able to cut her away from the rest of the band. He followed her for several hours. He pushed her. She tired, and he pushed her to unfamiliar water, let her tank up, and then roped her.

That afternoon, Korski took the paint up to a line cabin east of Needle Creek. Rita had told him that Andy, her husband, no longer had any use for the cabin and Korski was free to use it anytime. Call it your own, she told him the second time they'd slept together. Maybe that'll get you coming around to see me a little more often.

Korski liked the first part of what she'd said. He didn't much care for the second part. He'd told her the first time they went to bed that he wasn't one for commitments; he'd had enough of them. Rita knew almost nothing about him, only the name he'd given her, and his affection for her only son, Daryl. Lately, he'd been staying at the line cabin for several days at a time. He read novels. He hiked the nearby mountains. He hunted coyote, quail, and grouse, for food. He wrote in his journal.

Now he trimmed the paint's mane and tail and broke her. In no time at all, he had a halter on her and was teaching her basics. In the next three days, he rode her frequently. He finished Daryl's braided rawhide hackamore and worked more oil into the old Visalia stock saddle that he'd picked up on the 301 Ranch the first time he had ventured west of Ely into Long's Valley. Korski had used the saddle himself. He'd liked it so much that he doubted that he'd part with it. But he loved Daryl, would have wanted him as his own. Daryl already had a keen eye for the ways of Nevada buckaroos. Korski knew that Daryl would immediately recognize the saddle's center-fired, single-rigged cinch and the unswelled pommel with bucking rolls.

Daryl's present finished, Korski got antsy. He picked up a revised collection of essays by Montaigne. He read one called "Friends." Partway through the essay, he got a consuming urge for coyote steak. He picked up his .222, went outside and grabbed a handful of shells from his saddlebag, and headed out into the early-evening light. He circled behind the tiny cabin and started up toward the ridge. He knew exactly where to find a young and tender female.

Postmodern Conversations

[Postmodernism] is rich, open and contested.

—M. Doel and D. Matless, "Editorial: Geography and
Postmodernism," in *Environment and Planning D: Society
and Space*

SO academics have become postmodernists?

It's the rage. Their claims are considerable.

And what are they?

Many, and mighty—you wouldn't believe.

Tell me . . .

Your truth, my truth, the cricket's truth—no fundamental difference.

Don't get it.

Not sure I do either, but it goes something like this: Truths are socially conditioned, inherently unstable, creatures of the times. What you are is what I say you are.

And so your stories about me, yourself—whatever—send them to academic journals. If postmodernism is for real, then they won't care what you've sent them. Hell, they won't ask, won't think to ask!

True enough, I suppose. But there's more by way of running this postmodernist stuff to its logical conclusion. No more authorities. Which means no footnotes, no referees . . .

No judges.

And did I say, never again a question of appropriateness?

You think they'll wonder if there's anything geographical about this horse thief tale?

Impossible to predict with geegrogers. They'll call anything geographical if it suits their immediate needs.

You can whistle that one all day long.

I forgot to mention something important. You, the individual. Difference—one of these grand postmodernist concerns. Feminism, gender, sex, race, women, blacks . . . everybody's got to have their say these days. Even a white male like me must be given his say . . . I think.

So it's anarchy after all. Do what you do, call it what you will, and presto! It is, it lives, alongside the best, the worst.

You got it.

So you think they will—the professional academic journals—publish these stories of yours, or your alter ego?

Well, they're stuck, aren't they?

Do you think I really eat crickets?

A non sequitur.

Can't be, by these postmodernist rules.

Yes, true . . . but a distraction from this seriousness. If I send this in and they say it needs to be sent to referees, I'll say, Wait a minute, you're not taking this postmodernist thing seriously. Otherwise, why publish anything they have to say in the journals?

I get you.

There's more.

Go on.

Did I say there'd be no need for footnotes, references to other works, intros, conclusions. . . ?

Or need to hint or ever reveal if indeed you do eat crickets, steal horses, and know a delicious young girl by the name of Fifi?

Right.

And you'll now do exactly what with this story?

Get serious, test the waters—and why not? Like I said or did I say?, if I get a rejection slip, then I'll conclude that the only people taking postmodernists seriously are the self-described postmodernists.

And even they are in doubt, I imagine.

Exactly. Hey, imagine. This one just might be the first real shot in a revolution. I've got a file cabinet full of papers, musings, and what-have-you. Imagine all the academic journals that I can send the stuff to and look forward to being published without review, without worrying whether I'm telling truths or peddling dreams. Man, what a deal.

And you honestly think they'll publish it—this tale of horse thieving?

Well, who knows? It'll be the same bind for all of them.

And if they don't—if they're up to their old tricks of demanding foot-notes, using referees, and considering only standard kinds of articles and the like?

Then I guess we'll know, won't we?

An experiment worth undertaking, I'd say.

Costs me nothing. Just postage and a little time. The first gets paid by a university, and the second—well, I seem to have nothing but time on my hands these days.

Okay, where you going to send it then? Who are you going to embarrass first?

I'll start with the *Professional Geographer*. Seems like as good a place as any. And besides, no conflict of interest. Never met the editor, don't know what he looks like. Fact is, I don't know boo about him.

What're you going to do if he finds some clever way to reject this story?

Maybe I'll make the rounds of all the geography journals. Data points, you know.

That doesn't sound very postmodern to me.

You mean the bit about data points?

Exactly.

Whoever said I bought into this postmodern trip?

You haven't, then?

Hey, I can be converted—maybe. And besides, maybe I want to be con-verted. I've got that file cabinet full of stories I was telling you about. I bet I'm not the only one, either.

You look different.

Every day a little something changes.

The beard. That's what I'm referring to.

It's been a long time. It's got a ways to go yet too. Convergence—on you, you might say. I've been thinking about you a lot lately. Getting up there into the Limbo Range where they're going to burn our flesh and bury what's left when we're done with this sordid world.

Anytime, come along. Still got a camp there and a stash near the ridge line.

And you—you enjoying the raspberry dust again?

Don't like it. Too damn sweet. But you know me, can't get out of the sixties and those good head trips.

How come you're not giving me the good Gold anymore?

Haven't been able to get it. I haven't heard from the latest of my sweet little Colombian airline stewardesses in several months now.

Some bird in bed as I recall. Oops!—woman.

Doesn't matter what you call them anymore. Not unless you think this P.C. stuff is the same as the postmodern gig.

Well, is it?

Whatever feels good.

Hey . . . why are you bothering me today?

Got one of those data points we were talking about back on the seventeenth.

Already?

Yeah, me to him and back again in less than a week.

He must have thrown it back in the mail the same day he got it.

Could be.

What'd he say?

"I have received your manuscript and concluded that it is not appropriate material for the *Professional Geographer.*"

Man of many words, I see. Sounds like he might be protecting his ass.

They don't select for innovation or unpredictable take-chance sorts when they look for an editor that serves seven thousand tie-and-Gatorade sorts.

What'd you tell him in your cover letter? Maybe that was your undoing.

That he ought to take a careful look at it before he thought about rejecting it because it's a little something different. And that I didn't send an abstract.

What's this about an abstract?

Any old pretext will do you in. Didn't I ever tell you about the time Stanley Brunn, one of those know-it-all editors of the *Annals,* rejected one of my manuscripts because I didn't send an abstract?

You're pulling my chain.

Wish I was.

What else did you say in the cover letter?

Told him to send it to Allan Pred at Berkeley if he didn't understand what I was doing.

Who's he? Why him?

A run-hard geegroger who thinks he's a poet. Or some kind of newfangled philosopher, little hard to tell. Likes to arrange words on the page in funny ways, repeats himself a lot, uses the same word often—like Joan Didion.

Is he as good as Miss Migraine?

You kidding? She can *write*.

This Pred, is he postmodern?

Says not. But it's a little hard to say. Some other geegroger by the name of Michael Curry claims otherwise.

Maybe the editor called Pred, and Pred said, No way, this guy's just pulling your leg. He's famous for that. He's probably sitting in his office laughing himself into a fit over sending that to you.

I don't know. My guess is Pred would like to see what would happen. Get into the fray himself. Guy thrives on looking new. Besides, he'd be hypocritical saying no just because this true tale of horse thieving is different, looks like I was smoking dope when I wrote it.

Were you?

I don't remember. But it might be better if I said I was. Then Pred, whoever, could claim that my state of mind was part of the text. Then there'd be all the more reason for taking it. Hell, text is everywhere—another one of these postmodern ideas. Everything I've said, am saying, you, me, right now revealing I've got a half-empty beer about a foot away from my right hand sitting on top of the rejection letter from the *PG*.

Tell me about this editor of the *Professional Geographer*.

Like I said before, I don't know him from boo.

Maybe he didn't like you telling him where to send it. Sounds like something Fifi would do. By the way, how's she doing?

Working in a different house. She's now at the Lucky Strike Bar in Elko. Changed trucker boyfriends three times since I saw her last. But sassy and lovely as ever.

Give her my love.

Maybe, maybe not. I'm selfish about my lady friends.

You know, you could be right about the editor being uptight and all. This postmodern stuff is all clubby, inner-circle chatter. Got nothing to do with the main-line ordinary sorts that run all disciplines. Get a little off track, and they get lost and scared to death.

I read over what you wrote about me—you—and I think I might know one good reason he had to say what you sent was inappropriate for the journal. You had me thinking Fuck Gerbino when I saw that BLM bastard sitting up on the ridge looking at me through glasses, eager to haul my ass off to jail for stealing that paint.

Could be right, but that'd be just another measure of him not understanding what I sent, why it is so postmodern. If he understood, then he'd have to allow for who you are, whatever you said . . . and what's a little four-letter word, anyway?

Yeah, that's right. Remember the story about Norman Mailer?

Remind me.

About twenty-five years ago, maybe longer, he sent a piece of fiction to the *New Yorker*. Used the word *shit* in some dialogue. They said, Never, not in a thousand years in the *New Yorker*. He never sent them another piece. Too bad—it's long been the dullest magazine in the world.

But that was then, this is now. Such words are as common in the *New Yorker* as their cartoons. If it happened there, then it'll happen in geography journals, academic journals.

Dream on, baby. The academy's only half real, and that's giving it the benefit of the doubt. Retrogressive most of the time is more like it.

Well, then, why did you put that *fuck* in?

Because that's exactly what came into my mind the day I was actually stealing a paint and saw that BLM enforcer up on the ridge. And, anyway, how else am I going to test the limits of this postmodern jabber?

You could have put it in quotes.

Is it even in there?

Maybe. I need to go back and check, make another textual move.

Hey, it doesn't matter if it's there or not. That's actually what I thought that day about Gerbino back in '84 when I was after that paint. The thought was as real as I'm real. Anyway, real or not, anything goes now.

If it's not in there—your thought—then sneak it in after you get an editor to accept the story.

That's good. That'd be another good textual move. We could even dignify it by calling it a trope.

Maybe you should give this *PG* editor the benefit of the doubt about him being uptight. Maybe he's down-home, blue-collar, a teamster hitting on *fuck* every fourth word. Or consider other hypotheses.

I'm working at it. Right now I'm trying to get through to him on the phone. Called a few minutes ago, but he was in a faculty meeting. Maybe I'll call him again tomorrow morning.

Maybe you ought to just bury this stuff, go do something else.

Naw. Real friends like John of Oz, Jacquot, Don Pietro, and Quetzal would laugh if I gave up this easily, say it was out of character. Tell me I was getting soft, boring.

Where to next, then?

Think I'll try that journal Michael Dear was the editor of, one of those environment-and-planning journals. He published a boatload of postmodern chatter, not least his own. I'll tell the new editor to contact him to look at what I send.

What about Pred?

Tell him to contact him too.

How do you think Dear will respond?

Well, he's had some time to be thinking about it.

How's that?

I already sent him a copy of what I sent to the *PG.* Same day, matter of fact.

You didn't? I thought your favorite lady friend told you not to.

You talk to her?

Sure did.

I listen to her only some of the time. Besides, I wanted to tease Dear, get him thinking about what he's really saying.

You hear from him?

Not yet, but I ought to pretty soon. I invited him and his geegroger wife to come to Irvine. Said I'd take them out to a nice restaurant, a good fish joint over in Newport Beach.

I suppose you told him Korski was going to make reservations and he could talk to both of us.

Didn't go that far. Didn't want to rattle his cage. Besides, I said we wouldn't talk about postmodernism if they came down. Thought I'd be nice and polite.

That doesn't sound very postmodern to me, closing the door like that.

Hey, I only said I *might* be interested in being postmodern. First I want to be convinced that these journal editors, the discipline, are really serious about this newfangled stuff. I'm not yet sure who's listening to Dear and his buddies. No sense in proclaiming that I too have been born again if there's nothing to be born again for. Lots of costs involved in coming out postmodern.

What if Dear writes and says, You're off the wall, you scare the shit out of me, I don't know where you're coming from, and you're not playing by the rules?

I guess I'd have to say, What are the rules?

You mean he hasn't said yet?

Not even close. Probably plead there are no rules.

Then he'll love this stuff.

Ho, ho . . . He'll probably say, You read so-and-so Frenchman?

One of those kind, huh?

Says he's not into the authority trip, but he cites those obscure, impossible-to-understand Europeans like you wouldn't believe.

You think he'll still think you're nuts?

He might. But he won't be able to say it. Can't even admit it in public if he's going to give me the same space, voice, what-have-you he's so eager to give to everyone else.

You're going to do a few things different this time, aren't you?

What have you got in mind?

Changing the format to—

—whatever *Environment and Planning* uses? No way. I can't start diluting my strong hypothesis at this point.

Maybe you should expand on the story, tell everybody more. I mean, it gets better. Tell them more about Daryl and me, the paint, whether Daryl died, all about his mother, Rita, and how I was getting it on with her.

Not a bad idea, but right now I don't know how I feel about you and Rita, you sleeping with her every time her old man took off to Elko or Reno.

Hey, don't get into a moralizing trip.

Because that wouldn't be very postmodern?

Exactly.

Still . . .

Still, nothing. At least do me the favor of taking that *fuck* out of my mind. Forget me thinking about that prick Gerbino and the BLM trying to nail me on a federal felony charge when they ought to be congratulating me for getting another big, destructive mouth out of antelope country.

Could do, I suppose, but truth is that realist strain courses through me like I've got too many white blood cells.

Doesn't sound that way to me, the way you and I are carrying on trying to decide what to do with one of the most frequently used words in the English language.

Hey, look, Korski, why don't you piss off for the afternoon and let me get back to this dialogue on another story I'm trying to write?

All full of truths, no doubt.

For sure.

You call him?

Just like I said. Six o'clock this morning.

What'd he say?

Same thing he said in the letter. Like he didn't know what else to say, was scared shitless. Then I got him to mumble something about maybe there was something good in my last paragraph, but the story had nothing to do with it.

You explain it to him?

I didn't try on the story or what it had going for it, figured like most academics he wouldn't know story and good writing if they were slapping him silly. But I did try to give him the rationale for the story and the format. Three times with nine examples. Told him the last thing we need are more programmatic statements to the effect that we ought to do this or that thing. Said we need some honest-to-God examples, which is just what this tale about you is all about. Reminded him of all the empty-chatter geegrogers once had to put up with, people going on and on about existentialism, phenomenology, and humanism and never a concrete example.

So how'd he respond?

It was like talking to a stone wall. I gave him still more examples, told him we have to find out if fiction or its look-alike is legit, and the only way to do so is run a maybe-yes, maybe-no tale in front of our audience. Took him through a recent tiny book by Nick Entrikin called *The Betweenness of Place.* I said, Here is a book of fifty thousand words or so and a heavy-winded argument that we ought to be doing a narrative-like synthesis when we write about place. Not a single example in the book or in a follow-up.

Maybe this guy Entrikin doesn't know what he's talking about.

That's what I concluded after going through three volumes of Paul Ricoeur on *Time and Narrative,* more European gobbledygook that Entrikin genuflects to repeatedly in the wrong places when trying to convince us that there's something to this narrative-like synthesis idea that's his panacea for describing places.

But tell me more about this *PG* editor. Was he still a concrete wall after you kept after him with reason and examples?

No jackhammer I've ever seen was going to work.

You ask him if he sent it to Pred?

Sure did. Said no. Didn't send it out of his department, he said. Had it read in-house.

They got some postmodernists there, I take it?

He said, We got some people here who are aware of what's happening out there.

Where's this?

Some place in the North Carolina sticks, can't recall exactly where at the moment.

You ask him how he handled looking at *fuck* on the page?

Didn't want him hanging up on me.

But you kept pressing him anyway?

You know me, I always go full court. Told him to try Pred, Dear, Curry, I'd give him a dozen names of other people to contact if he didn't want to go where I told him to go.

And?

He repeated what he had said when he first opened his mouth. Like he was hardwired into a Rajneesh mantra. So I thought about it a minute and then thought, Aw, fuck it, I'll need dynamite, and I'm not sure that'll work either.

You let him know you were pissed?

There are other ways.

After I talked to the editor at the *Professional Geographer*, had a little breakfast, took my three-year-old boy, Cole, to school, had another talk with Korski, and put it all down for posterity, I decided, Hell, I think I'll go down to the Imperial Valley for a couple of days and snoop around and see if I can get some more insight into how I'm going to argue that we ought to just go down there and take all that Colorado water so we can have enough for all these Mexicans pouring across the border and also feed my habit of growing black and golden and weeping Mexican bamboo all over our yard, nothing I'm ever honest enough to tell the thousands of students at the university I preach to about saving water, thinking of future generations, and finding ways to depopulate this part of the state.

By noon, I was up in the Cleveland National Forest, standing by the side of the road, staring at small patches of snow, and thinking about that Laguna Beach fire and how I was going to handle some not-so-small discrepancies between what my eyes told me I saw three days after the fire and what fourteen reporters (their names all there on the front page of the Orange County edition of the *L.A. Times*) were authoritatively telling the rest of the world, about the fire destroying trailers on El Moro beach and, in all, burning one hundred of them in the trailer park, a problem because my eyes told me that the number did not exceed forty, all forty of them on the east side of the Pacific Coast Highway, and there wasn't a single trailer on the beach, a couple of hundred yards away and across the highway, that got more than a little soot on the roof or porch. I thought about this detail, not least because that same Stanley Brunn I told Korski about asked me for footnotes for statements based on fieldwork on horses in a paper that I sent to him about a year ago, and I was so dumbfounded by his request that I thought, Christ, when was this guy born? And then I thought, Well, at least Michael Dear might've come to my rescue, saying, Hey, Symanski's got an opinion on the matter too. True postmodern democracy.

I got back in my pickup and stepped on the gas, and I didn't stop until I got to the decrepit Signal Mountain restaurant west of Calexico where I met the one-time Mexican bullfighter, a lady wearing a gorgeous red cape and sporting a black, wiry mustache the likes of which I have never seen in all of my travels, which, when added to the conversation that followed, itself an otherworldly mix of English and her singsong Mexican Spanish, got my day rolling on one surreal trip—ending, or almost, with me chasing the ghost of a very old girlfriend that still makes me shudder in the most pleasant sort of way, she was that good in the rack. And this encounter came before I even thought of buying a pint of Johnny Walker Black and going to the last stick of Santa Marta Gold in my possession.

There were three people in that Signal Mountain restaurant, and me. One was an Anglo sitting near the front door eating tacos and reading the newspaper; he had a white knot on the right side of his nose as big as my thumb, and a dewlap that a Paiute would say would make one hell of a water bag. One was the bartender, a slave from Mexicali, I guessed, who didn't know a word of English and seemed puzzled why I had all these questions about the fifty-odd bullfighting paintings and posters on the walls and didn't have the courtesy to order a beer or ask for the lunch menu. And then there was Doña Maria, sitting there on the north side of the restaurant in a booth just outside the kitchen, beneath, I would discover, an arresting picture of herself in full matador regalia at the age of fifty, this picture not far from one of her Basque husband, now fifteen years dead, who, significantly, opened the restaurant in 1950.

The conversation with Doña Maria had more layers of meaning than I can begin to describe. But the telling kernel of truth was this number fifty that kept coming up, first in when the restaurant was opened, and then in her telling me that the picture of her looking as lovely as a fresh, young JAP was taken when she was fifty, and then her telling me, all evidence to the contrary so said my one good eye, that she had fought her last bull in Mexico City fifty years ago. I circled the number fifty maybe a dozen times, looking for the trick in repetition, trying to conclude and not wanting to that it was all coincidence and I shouldn't make much of it, no more than Kundera makes of coincidences in *The Unbearable Lightness of Being.* Finally, I just had to excuse myself, and I left Doña Maria. I backed away from her slowly, as if she had a gun and might shoot me in the back if I turned. This, of course, made me feel most uncomfortable and not at all myself, as I cannot recall the last time I didn't take my chances, even that night when I beat the shit out of that biker in the alley back of the Alley Cat after he'd tried to steal my swag. Yeah, I'll admit it, Doña Maria and that number fifty had me rattled.

But then by midafternoon, after rolling around the scummy streets of Calexico looking for oddball impressions and some decent graffiti and wondering what had gotten half the people on Imperial Highway to paint their businesses in two colors—red and yellow—I got a hotel room, or rather negotiated one with a feisty Mexican girl with some beautiful teeth who couldn't decide where she was born and who was most unhappy with me when I said I wouldn't take the twenty-five-dollar room until she put a new lightbulb in the bedside lamp, explaining to her that among my many addictions is the habit of reading late at night, and that on this particular night I expected to be reading some engrossing words about wandering in the desert. That's another addiction of mine—deserts.

The best I could find in bar-poor Calexico was the Calexico 111, a small, dark joint with two pool tables, one TV that was showing Purdue and Michigan proving to each other that neither was good enough for the Final Four, and two Mexican guitar players soliciting the half-dozen Mexicans at the bar to see what kind of music they pined for. I had only one beer, my first of the night, and was ready to leave when I looked out into the dark parking lot at the rear and saw a couple of Mexicans down on their haunches in front of some pickups. I picked up just enough light from the rear of the bar to see their dice in the dirt. If there wasn't going to be any action in the bar, I figured a little dice would make up for it, particularly when I got out there and saw one of the Mexicans with a wad of bills that was as fat as a dictionary.

After about five minutes of seeing what they were up to and what the stakes were, I changed in several twenties for a stack of ones and fives and got in the game and started heating up for it. I lost a little and won a little, and then suddenly I heard this shuffling noise in the alley. I looked up and at first ignored what I saw. And then I couldn't, I just couldn't, for I was certain it was this girl of long, long ago that still makes me hard when I think of all she could do to give pleasure. She looked, I would swear, exactly as I remember her from those graduate school days in the middle of winter when I walked her home from the Vulgate at three in the morning through slush, calf-high snow, and biting-cold temperatures to find myself inside her two-room apartment ripping her clothes off and her doing the same to me.

It was Blue, or the ghost of her; it had to be: the long black hair (of course, any color might have been black in the alley), the right height—about five ten, the skirt about two inches above the knees, that tiny nose, and those high cheekbones in silhouette—absolutely perfect. And, then, the most unlikely thing to be seen anywhere in the whole of the Imperial Valley this time of year, the boots that came just below her knees, and the finishing touch—the socks that ended just above the boots, just below the knees. I would not, could not, have been so certain were it not for that alley light under which she walked as I slowly stood and stared, unable to believe my eyes, starting now to shiver at the image of her, fearful for what it might mean, suddenly now bringing to mind Doña Maria.

Foolishly, I dropped back down on my haunches and put some bills in the pool, took the dice, and threw them. And maybe I threw them again, I cannot now recall, such was the disruption to my mind in the minutes following. I do recall feeling suddenly cold, almost icy, and then I stood. I peered over the hoods of a couple of pickups, in search of this girl—girl, hell!—that I would never forget. But Blue, my Blue, was gone! I remember yelling, at the top of my lungs, Shit! And then without even saying, See you later or Another time, I stuffed the money in hand in my shirt pocket and scurried into the alley in search of her.

Nowhere. I just couldn't see her. And this seemed impossible.

I knew exactly which direction Blue was heading, down toward the hardware store and the Chinese restaurant, and the spot where, earlier in the day, I had seen a Mexican bum pushing a shopping cart and sticking his nose under parked cars in search of beer cans, bottles, and cigarette butts. So I ran, turning my head to the left, then to the right, stopping, starting, looking to my rear, cussing. Where could she have gone? I wondered. This is impossible! I found myself shouting at the stars.

I backtracked. I took a side street, then another.

And then I found myself dodging honking cars on the Imperial Highway, almost getting run over, one guy yelling, Hey, asshole, watch where you are going.

Oh, shit, I thought, knowing I'd never again be able to enjoy all that I had truly enjoyed with my lovely, voracious, one-of-a-kind Blue.

Part Two ◎ Worthless Words in the World

Three years out of the field, I think I realized that I didn't want to be the anthropologist but the informant.

—N. Tarn, "Interview with Gary Snyder"

Richard at nine

Calling a Spade a Spade

If you are looking for the gravy, young man, it's right under your nose.

—David Lodge, *Small World*

A semester prior to taking a graduate seminar from D. W. Meinig on Anglo America, I took a couple of graduate courses in the philosophy department at Syracuse University. In one of these classes, a good deal of attention was devoted to the latter Wittgenstein, especially *Philosophical Investigations.* Like many who have read and tried to deal with Wittgenstein, I became fascinated as much by what he had to say as by how he said what he did. What I picked up from this course, and from close friends who were graduate students in the philosophy department, was very much on my mind when I enrolled in Meinig's seminar.

Meinig gave students the freedom to pursue whatever topic caught their fancy. So I told him of my fascination with Wittgenstein and his preoccupation with language puzzles, and that I wanted to see whether such a concern led anywhere fruitful. Would he mind if I examined his own work, since it was well known that he was admired for his felicitous writing style? He said no, he would be flattered.

At the time I made this proposal to Meinig, I said that I did not know where such an analysis of some of his writing would lead, that I would know only after doing it. He seemed unconcerned, confident, I think, that nothing I might do would impugn either his style or his scholarship.

My seminar paper for Meinig that semester analyzed only a couple of paragraphs from his small book *Imperial Texas*, paragraphs that I claimed were representative of the book. My analysis ran to roughly four thousand words. I reached conclusions similar to what I would reach in "The Manipulation of Ordinary Language."[1]

When I finished presenting the paper in the seminar, Meinig stood and said that he had been "numbed" only twice in his life. The first time he had a kidney problem that required an operation. The second time, he said, had just occurred—and from the waist up—as I presented my analysis and conclusions regarding his writing style and truth telling. He said nothing more.

The next week, Meinig invited me to his office to discuss the paper. The discussion was brief. He asked only a question or two. He did not take exception to what I had concluded. He gave me an A in the course.

I more or less forgot about Meinig's use of language until the fall of 1975 when the Department of Geography at the University of Minnesota asked me to participate in its Distinguished Lecture Series. They wanted me to give two formal presentations. At the time, I had been giving thought to the law of natural selection and how it might be relevant to geographic inquiry. I would present one paper on this topic. I had no desire to rehash findings from my work on periodic markets in Colombia, because they were already in print or forthcoming in journals easily accessible to geographers.[2] I thought that the most interesting material that I had not widely shared with others was my one brief foray into ordinary language analysis—the Meinig effort. When I looked over the seminar paper I had written, now some seven years behind me, and then reread Meinig's books, I concluded that the basic approach and conclusions reached were still valid. But I would need to do additional analyses for a paper good enough for a Minnesota lecture.

Aware that some who heard me at Minnesota might—despite disclaimers to the contrary—think that I was personally attacking Meinig, I asked Meinig if he objected to my sharing my views on his writing and scholarship with the faculty and students at Minnesota. I also promised to send him a copy of what I intended to present and added that I would be happy to share his criticisms of my paper with the audience. (I was, of course, under no academic or other obligation to do so.) Meinig gave me his blessing for presenting the paper, I sent the manuscript to him as promised, he provided a brief critique, and I read it after presenting the paper at Minnesota.

Upon my return to the University of Texas at Austin, where I was an assistant professor of geography and Latin American studies, Paul English, the de facto head of the department, invited me to lunch. Thereupon, he verbally berated me for having the audacity to criticize anything that Meinig had written; I would, he told me, "pay" for this mighty unforgivable misstep. In a glowing review of one of Meinig's books, English has described Meinig as a rare geographer, someone clearly in the belles-lettres tradition. Paul English had neither seen nor heard what I had to say about Meinig, and his only information on my presentation at Minnesota came from a telephone conversation (English told me) with one Minnesota faculty member (Ward Barrett was the name he gave me).

Paul English was not joking. Within a year, he personally engineered a departmental vote to get me fired. The seven-to-zero departmental vote was unanimously overturned by the College of Arts and Sciences.

In the spring of 1976, the Minnesota lectures behind me, I did further analysis on Meinig's writing, shared the paper with several people (not including Paul English), incorporated criticisms that seemed appropriate, and then submitted the manuscript to the *Annals of the Association of American Geographers*. It was a matter of only weeks before I heard from the editor, John Hudson. He informed me that he was accepting the paper with minor revisions. The turnaround time, by academic journal standards, was phenomenally short. The paper appeared in December 1976.

Either during the process of revision or not long thereafter—my curiosity about the short "review" time now nagging—I asked John Hudson how many people had reviewed the paper. He said that only he had reviewed it, that it was not necessary to send it out for peer review because the people whom I acknowledged in the paper were the only ones he could think of who were qualified to judge its scholarly merits. It was obvious nonsense on his part. I assumed that were he questioned on the matter, he had one particular person in mind whom I had acknowledged: Gunnar Olsson, then of the University of Michigan.

I concluded that John Hudson published the paper unreviewed and with haste for two reasons. He more or less agreed with the analysis. And because of his intellectual shift away from diffusion and mindless mathematical modeling of various sorts to a commitment to narrative historical geography, somewhat akin to the kind done by Meinig (though with greater attention to "hard" data, to Hudson's considerable credit) on Anglo America, his own future reputation would be tied to Meinig's. If Meinig looked "bad" or less estimable than most geographers thought, then he—Hudson—would look good by comparison.

I have long thought that John Hudson could not easily have found two or three people in the discipline to support "The Manipulation of Ordinary Language" with enough enthusiasm to strongly recommend publication in the *Annals*. Most academic geographers would have wanted to protect Meinig's reputation, and they would have looked for every imaginable reason for doing so. To this end, I am sure, they would have been quite willing to suppress the manuscript (much as they would have been eager to suppress "Prostitution in Nevada" were it peer reviewed—which it was not),[3] even if they believed that its conclusions were sound and the paper merited publication.

Subsequent to the publication of "Manipulation," I got a lot of flack. Despite my statement in the paper that anything I said should not be construed as a personal attack on Meinig ("Lest my purpose be misunderstood, I should emphasize at this point that the following analysis is not a personal attack on Meinig; it is his work, and only his work, which is the subject of concern"), the people who wrote or spoke to me about the paper said that I *must* have had a grudge against Meinig to have written what I did.[4] I would not, these critics argued, have otherwise done so.

In fact, up to the time of publication of "The Manipulation of Ordinary Language," I considered Meinig as much a friend or ally as any of the faculty at Syracuse, notwithstanding my unhappiness over his cowardly handling of a moral and ethical issue I'd been involved with as a graduate student. The reasons for initially analyzing Meinig's writing were straightforward. I had a strong interest in language usage, Meinig had a very high profile within the discipline of geography for his "superb" prose, and he was close to home—literally down the hall and around the corner from my graduate student carrel.

Both in writing and in one-on-one conversations, I openly invited geographers who expressed concern about "The Manipulation of Ordinary Language" to critique my position on Meinig's writing. I told them to send their comments to the *Annals* for publication or send them to me and I would respond. With the exception of my Ph.D. adviser, a grade school teacher in Wisconsin and a historical geographer at a Canadian university, no one had any interest in my offers or encouragement.[5]

Off and on in the years after the publication of "The Manipulation of Ordinary Language," I would hear through fairly reliable sources that John Hudson had received nasty comments or notes because he had the audacity to publish the paper, and that this or that person—often named for my benefit if I "promised not to tell or confront the person"—"knew" that my motives for writing the paper were perverse, malicious, or downright dishonest. They were invariably people whom I had never met or even known through correspondence.

Now others beside Paul English at the University of Texas at Austin let me know that it had been "uncouth" and "unprofessional" for me to have written what I did about Meinig, just as it had been similarly "out of bounds" and "unprofessional" for me to have written "Prostitution in Nevada." No specific critiques of either of these papers were ever offered by my University of Texas colleagues.

In my battle with the Department of Geography at Texas concerning my retention on the faculty—an issue very much related to publishing "Prostitution in Nevada" and "The Manipulation of Ordinary Language"— I asked Meinig for a letter of support. The support he gave, given my publication record, was, at best, left-handed. In his letter, he drew attention to what I had written about his writing, and in a single sentence dismissed it as misguided and wrongheaded.

Although I had sent my Ph.D. adviser, David Sopher, a copy of the manuscript submitted to the *Annals* months before final revisions were sent to John Hudson with a plea that he give me as much feedback as possible, he gave me none. But then after the paper was published, in person and in correspondence, Sopher viciously attacked me and my motives. Though he had no evidence to offer, he was certain that I "had it in for Meinig." He would hear nothing of my protestations or my wife's (girlfriend at the time the paper was written), and he would not believe that it was only Meinig's work and not Meinig the person that concerned me. He was unmoved after a long monologue by me on the absolute centrality of and need for dissent and critical posturing in the academy. Sopher offered neither excuse nor apology for not having read the prepublication manuscript I had sent him.

From all these events, I concluded that a fair number of academic geographers came to several conclusions about this paper and me: notwithstanding whatever merits my analysis of Meinig's writing may have had, I obviously had a grudge against Meinig, and writing that paper was my way of settling a score. The proof, I was repeatedly reminded, was that I had been a graduate student in the same department where Meinig held a distinguished professorial chair. Thus, propinquity and malice are as brother and sister, cause and obvious effect. But even if I did not exactly have it in for Meinig, I, as a young assistant professor, had no "right" to "attack" someone as esteemed and well regarded within geography as Meinig clearly was. The issue here was not at all about the merits of an intellectual argument, but about my inferior position in a social hierarchy. Dissent in the academy, in this view, is permissible only where kings are judging kings or kings their vassals. Serfs, slaves, and assistant professors are without rights; they are, as my father often said to me when telling me to get lost in the presence of adult company, to be seen and not heard.

Then too, it was now easy to conclude that I was not merely a sinner who could be forgiven a one-time transgression, but rather someone who obviously delighted in stepping beyond the bounds of what tribal norms dictated were normal and acceptable behavior. Lest anyone doubt this conclusion, "Prostitution in Nevada" was the other piece of incontrovertible evidence that I was hopelessly beyond salvation.

The last time I saw David Sopher, someone for whom I had had a great deal of affection during my graduate school days, it was clear that my work both on prostitution and especially on Meinig's writing had created an emotional and intellectual Grand Canyon between us. Sopher had come to see me as his black sheep, an advisee to be disowned. It became apparent when my wife and I went to visit David and his wife at their Syracuse home in the chilly fall of 1977. The experience was wrenching, and, though I did not know it at the time, it would be the last time I would ever see him. A couple of years later, I heard through a friend that David Sopher had died of prostate cancer.

Now, more than thirty years after I first became interested in how language is used and how it is used by Meinig, I have again returned to his work, this time to a huge effort that when finished will have consumed more than twenty years of his academic life (see the chapter "Ruminations on a Misshapen America"). I did not expect that he would pay any attention to what I wrote about him in 1976, and it is pretty clear that he, and others, did not. With a fresh look at his work—or rather the first two volumes of a four-volume work—I find even more reason to be unhappy with both Meinig's writing and his narrow approach to geography. I quite expect that this critique will be received even less kindly than the previous one. I now hold no position in academic geography, Meinig is much "larger" academically than he was twenty-five years ago, and my outland reputation is, building on the labels I began to acquire in the 1970s, even more secure.

Deconstructing a Drugstore Cowboy

> Words are the counters of wise men, and the money of fools.
>
> —Thomas Hobbes

PAUL Starrs, editor of the *Geographical Review,* is in love with words. At times he can string them together with the deftness of a seasoned buckaroo knifing and moving Rocky Mountain oysters from bull to mouth without exhaling. He's got a sizable and offbeat vocabulary, and an equally queer sense of how to manipulate grammar in a way that catches the eye of even the most seasoned writer. Now and again, if not too hot to handle, he's even eminently quotable, anything but ordinary fare. You might even leave a Paul Starrs meal feeling a bit giddy. But then, after sobering up, you feel positively confused. Alas, the confusion does not slacken, and you begin to wonder exactly what he was trying to say.

This love of words and language comes at a cost, and a fairly pricey one at that. Because words and ways of saying are so prominent on his list of priorities, Starrs consistently plays down or ignores analysis: the tight logical flow of an argument, the marshaling of evidence, the bonding glue between argument and evidence. Intellectually, then, as scholar or scientist, Paul Starrs is full of flab, a purveyor of nonexistent connections or misshapen ones, a believer in the irrelevancy of such matters. His intellectual forefather is not his mentor, Jim Parsons—the long-time Berkeley Latin Americanist, student of the American West, and one of the better writers in academic geography in the twentieth century—but the hustling essayist, the provocateur, the underage Mencken on the make.[1] And like Mencken at his worst, and sometimes when he seemed to be at his best, Starrs, in writing mode, can be full of cheap shots, one- or two-liners ready for the coffin the very afternoon they were spawned.

Take, for example, his book *Let the Cowboy Ride.*[2] This book is as good an example as any of what makes you want to read him, hate yourself for having taken him seriously in the first flush of the first reading, and then want to let it be known, to him and his admirers, that eating those range-grown oysters raw is a bad idea indeed.

Anyone who has spent any time in the desert West—the basin and range country of Nevada (Starrs's adopted home and where he teaches in Reno), the southeastern canyon expanses of Oregon and the Owyhee, the immense walled valleys of far northeastern California, the Red Desert of Wyoming, a good hunk of Colorado, and most of New Mexico and Arizona—knows that a big issue, and one impossible to miss, is overgrazing of fragile desert ecosystems. Sheep, cattle, horses (wild ones), and donkeys (also wild) love to eat most of what's out there, and they'll find whatever water's around and do whatever needs to be done to get to it. With their big bodies and trampling feet, they can make an awful mess of an awful lot of this vast part of this magnificent country very quickly. And they have—there can be no doubt. They've done so historically, and they've done so right down to the present. The federal government, in particular the Interior Department's Bureau of Land Management, knows all this, the ranchers who get to use public lands for pocket change know this, environmentalists who have trekked even spottily this country know this, and so does Paul Starrs. He knows not just how important the issue is in any assessment of cowboy culture and western ranching, but also that in a great many places (to be sure, not everywhere), the damage to irrecoverable soils, native plant life, and nearly extinct species is considerable.

I know it too, and not just as someone who has read a big hunk of the same literature as Starrs. In the early 1980s, I worked in a number of western states, mostly in Nevada, talking to ranchers and a lot of BLM personnel about wild horses, cattle, and sheep, trying to get some sense of who or what was doing all the evident damage to riparian areas and open ranges. I hiked and rode over and through a lot of desert lands used and abused by cattle, horses, and sheep. I examined piles of published government documents, and, by simply asking for access in district BLM offices, I got inside the actual grazing and permit files of dozens of ranchers. There I could find out exactly how much grazing land they had access to, how many trespasses they had received, how much they had paid over the years for abusing their rights to public lands, and how good a case the government had for lifting their grazing permits. This information in hand, more than once I accompanied BLM range specialists on transects to measure grazing pressure and abuse. Then I would sit with BLM managers and range specialists in Tonopah, Reno, Winnemucca, and Elko and get specific about this or that rancher: why he was running so many cows; why the feral horses roaming free hadn't been shot or rounded up; and when the BLM would find the courage to put a noose around the neck of certain ranchers and put some out of business permanently, reduce the size of the grazing permits of others, and take some land out of use for the foreseeable future. Invariably, the issue was not whether there was overgrazing, but whether it was politically feasible to go after ranchers with power all out of proportion to their numbers or contribution to any economy.

It is perhaps worth beginning a look at Starrs on grazing in an otherwise unusual place, in his twenty-page "Bibliographic Essay" that precedes the index. When we look in the index under the heading "overgrazing," it turns out that Starrs has almost as much to say in the bibliographic essay about grazing as he does anywhere in the book. And the way he handles this serious grazing matter, absolutely central to all politically and environmentally aware discussions of the cowboy West, is genuinely revealing.

Under a section in the essay titled "Images," Starrs tells us about two "remarkable" studies, Kendall L. Johnson's *Rangeland Through Time: A Photographic Study of Vegetation Change in Wyoming, 1870–1896* and Gary F. Rogers's *Then and Now: A Photographic History of Vegetation Change in the Central Great Basin Desert.* We're told that the authors "did actual measurements of forage, quality, and cover using paired historical and contemporary photographs."[3] Yet Starrs's use of these findings in detailing and noting regional differences, the specifics of ecological change, and what it all says about responsible grazing or overgrazing is practically nil. He just doesn't want to address the grazing issue, or address it seriously.

This lack of seriousness—to the point of utter irresponsibility—is fittingly highlighted on the following page of this essay in a section titled "Opponents." Here, rather than in the text, we are given an "analysis" of a book on grazing that needs to be taken seriously. The book is Denzel and Nancy Ferguson's *Sacred Cows at the Public Trough.* To Starrs, however, the book is nothing more than an "antigrazing screed."[4] *This statement* is Paul Starrs's "analysis" of the book.

On another book that takes issue with grazing practices in the West, Lynn R. Jacobs's *Waste of the West: Public Lands Ranching,* Starrs again gives us no analysis here or in the text. He makes no attempt whatsoever to explore the issues and show that the author of this book has arguments that are silly, possess merit, or are worthy up to a point but in need of qualification. Instead, *Waste,* in one sentence, is reduced to the charge that the author is "the current champion of bovine hating" and that the book is "a kind of monument to distaste."[5] Yes, it is true that the book leaves something to be desired in the way it is written and formatted, but the author in fact has a good deal to say that ought to be seriously addressed, not simply written off with a pair of eye-catching statements that convey no information at all.

Others who have been unhappy with the unconscionable treatment of the public's (not the ranchers') land also get quickly whacked, not least just about everybody's favorite target, Jeremy Rifkin, this time for his book *Beyond Beef: The Rise and Fall of the Cattle Culture.*[6]

All the bad guys dead and ready for the frontier mortuary wagon, Starrs now lets it be known that people like himself who are standing tall with smoking guns on what he describes as this "absolutely bimodal" landscape know that there's really a "middle ground" where reason prevails. Here, we can surmise, we've finally found Gary Cooper and Gregory Peck at their best: people with a "remarkably intelligent sensibility," with a "steadying hand," thank God not "alarming" and "inconsistent" like the ones in the "green" movement.

To reiterate, what's needed here, in a text with notes that runs to 321 pages, is a careful analysis of the issues: a listing of the charges and countercharges, the evidence on both sides, the case for present as opposed to past overgrazing as opposed to long-ongoing grazing, the case for overgrazing by horses as opposed to cattle and regional variations of these obvious culprits, and the evidence—of which there's a lot around—of what has happened to riparian areas, and the bird and fish populations that used to be found along and in creeks and streams.

That Starrs is simply lost on this grazing issue and that there is indeed a complex picture to be discussed and unraveled region by region and through time are nicely illustrated by a book that appeared one year after Starrs's *Let the Cowboy Ride.* The book, Debra Donahue's *Western Range Revisited: Removing Livestock from Public Lands to Conserve Native Biodiversity,* is so superior to Starrs's effort—yet makes no claim to be comprehensive on the grazing issue—that the one is clearly the work of a professional, the other the work of little more than an undergraduate neophyte.[7]

What's then needed by Starrs is, at the very least, a large summing up on the order of what Donahue has done, a reasoned conclusion drawing on evidence that an intelligent reader can identify with. What is not needed is yet more evidence from Starrs that he dances with words like he may well dance all through the night at an Elko redneck rodeo bar.

I've examined in some detail what Starrs had to say on two of the five pages under the index heading of "overgrazing." For a third page to which a student interested in the topic is directed, we're given a page number that takes us to the footnotes. That footnote turns out to be about that "rare" case "where abuse of public rangelands is a real problem."[8] Starrs here brings to light the case of John Casey (whom, I should note, just about everyone has heard about who has sniffed around in Nevada for a week or so) as "quite possibly the last of an older breed" of land abuser. This statement is just plain, pure distortion. Casey—bad guy that he is—is by no means the last of any breed, old or new, and if Starrs had done his fieldwork in Elko or any other number of BLM offices, then he would know as much. And—maybe this fact is really telling—he wouldn't have had to cite a newspaper, the *Sacramento Bee,* for "documentation" on Casey. For someone who lives just half an hour down the road from all kinds of overgrazing, and the state's major BLM office, the last person Starrs should go to for evidence of any sort would be a journalist, to boot one who lives at least a couple of hours away in another state.

When we turn to the text itself—only two pages devoted to overgrazing (so says Starrs's indexing effort)—we find that on one page, Starrs wants to embrace a refrain widely voiced by ranchers, namely, that any evidence for overgrazing is historical. In Starrs's own words: "The years from 1880 to the mid-1930s marked a half-century of destructive exploitation of western rangeland that has few North American equivalents for intensity of rapine or unregulated voraciousness."[9] Nothing in this charge is explored, elaborated, or documented; assertion is good enough for Starrs. Here too, on this page, there is virtually no sense whatsoever that ranchers have been anything but good and responsible stewards for the past sixty or seventy years. And the people who would take a contrary position are, to Starrs, "militant elitists" and "disaffected." Again, he gives no analysis or sustained, or even partial, argument as to why the ones opposed in whole or in part to grazing on public lands are elitists or, worse, I guess, disaffected elitists.

On the fifth and final page to which a reader interested in overgrazing is directed, there is absolutely nothing on overgrazing, nothing more than some vague statements that there has been tension between ranchers and the federal government.

When we turn to other possible index entries where we might hope to find more on the overgrazing issue, namely, "abuse" and "exploitation," we find that they are absent. If we then turn to the entry "wild horses," there is but a single page number given, and it is, again, to a footnote. And that footnote, as it happens, draws attention to my book *Wild Horses and Sacred Cows*. But rather than explore the many issues around wild horses and the Wild Horse and Burro Act of 1971 that I examined, Starrs gives us nothing more than the gratuitous and utterly hyperbolic comment that the Wild Horse Act is "among the most grotesque laws in American history, guaranteeing on-going and protected status for exotic animals that are enormously damaging in their foraging." Damaging they are indeed, as I have written in *Wild Horses*. But the actions of the BLM—and how it has managed the Wild Horse Act and aided or slowed damage done by horses, cattle, and sheep—are much too complex to be reduced to one sentence in a footnote and two assertions in the text, one of which ends with the footnote to my book. The two sentences read: "Few thinking westerners would deny that there has been considerable damage done to waterways, vegetation, and the general physique of the American West by the grazing of introduced herbivores (wild horses and burros, however, are untouchable). How much is acceptable, and how much the depredations of grazing compare with those of other activities, is more a policy question than a biological absolute." [10]

Starrs is just dead wrong about westerners "knowing" about all this considerable damage done by wild horses and burros. He is dead wrong that the wild horses and burros are "untouchable." The BLM is rounding them up all the time to control their numbers. And Starrs is thinking with his feet when he writes that when it comes to grazing, it "is more a policy question than a biological absolute." This assertion not only makes no sense, but is just dead wrong, again.

In an extended discussion of Elko County, Nevada, and the issue of grazing, Starrs says the following: "[A]nalyses of contemporary travel accounts and rephotography—examining nineteenth—or early-twentieth-century photographs by taking contemporary shots of precisely the same site and analyzing the degree of change—have suggested that the landscape was not *massively* different, visually, from today." [11] Again, Starrs wants to place the blame for the poor condition of Nevada's rangelands on all those long-dead cowboys. Their offspring have, as Starrs would have it, done no worse than maintain matters as they were more than half a century ago. But Starrs has made no effort whatsoever to walk the ground, alone or with others from the Elko BLM office (and they would gladly do so with him or me or anyone for the asking), to see if in fact what the range looks like today compares to what it looked like a hundred years ago, using photographs that, I suspect, reveal a whole lot less than Starrs claims or wants to believe. Once again, Starrs is not doing anything like good science or even merely decent journeyman analysis. He is just playing the unconvincing game of making repeated assertions, with the hope that the mantralike exercise will numb the inattentive reader into believing what Starrs himself wants to believe, not what he has shown, or wants to show, by way of fieldwork data or a close examination of what others have had to say about the matter either in Elko or anywhere else.

Three pages from the end of the text proper, Starrs tells us that he has worked for seven years on ranches in the West, and then, a paragraph later, he tells us that "Americans have become trendier and credulous, more sophists but even less analytical and commonsensical, since World War II." All of it is by way of trying to tell us that he's not a sophist and he is good at analysis and uses a lot of common sense, but also saying, in effect, Believe me, the cowboy way of life and ranching in the West are more complex and more enviable than you, the reader, who has not experienced it firsthand as I have, can possibly know, and therefore, let me remind you that "however the antigrazing forces might attempt to paint the picture," they've just gotten it all wrong.[12]

Notwithstanding some concessions he's willing to make here and there about ranchers not always being responsible, Starrs is just a polemicist. He is not, in this book, and on the very important issue of grazing and use of public lands, to be trusted. Indeed, he and this kind of book are dangerous. The book is dangerous because it has all the wrapping of first-rate scholarship, because it was published by one of this country's best academic presses, and because Starrs has badly misrepresented and lost sight of the grazing issue by encasing it in a lot of fairly fascinating material that makes it seem as though he has pretty much exhausted what's worth saying about cowboy country.

There's more to be said, and more pages and paragraphs I might slowly move through, but the point I'm making is, by now, fairly obvious. So it is worth turning, if only briefly, to the issue of why Starrs has treated this serious overgrazing matter the way he has.

As I've already suggested, one reason is that he's completely caught up in language, and words. Responsible scholarship, and analysis that's worth calling analysis, is of secondary importance. Or none whatsoever. Not only does Starrs not demonstrate any analytic abilities on the grazing issue, but based on the evidence, it's also questionable whether he understands what responsible scholarly analysis is all about.

Like many a drugstore cowboy, Starrs is caught up in his own largely imaginary rangeland-lassoing past (no doubt much exaggerated), the romanticism of the cowboy and the Old West. It's evident throughout the book, and it's most tellingly evident in his personalized University of Nevada website.[13] The glaring showpiece of his home page is a large color photograph of himself (at a younger and trimmer age) in seasoned chaps and well-traveled cowboy hat and boots standing alongside a saddled horse.

Although on this same website he touts "cultural aspects of research management" and "pastoralism (land use, animals, herders & cowhands)" among an astonishing range of research interests (two to three times as many as I can imagine anyone seriously dabbling in), it's pretty clear from the course he teaches called "Nevada: Patterns on the Land" (the details are on his website) that he gives no more serious attention to the real nature of grazing practices in what he tells naïve students than he does in what he has written in this book. From all the evidence, it appears that he just flatly refuses to allow himself to see the state's ranchers, among the worst in the West, as the voracious land abusers that they surely are. Paul Starrs is living a love affair in an Old West dreamland. He is, in his own way, as ideologically conditioned and driven—and as blind by design—as any Marxist, born-again Christian or Islamic fanatic.

Starrs's own prejudices are reinforced and magnified by teaching in a university, the University of Nevada at Reno, that is far from one of the better places in North America for taking a coldly analytical eye to western grazing practices. I know this fact from having talked to professors who teach there; there are credible stories about faculty who have wanted to take a close look at what's going on with ranchers in the state, and it has not been easy. Yet I strongly suspect that Starrs's shoddy handling of the grazing issue would not have looked much different if he were at a distant university when writing about cowboys and grazing issues; he came to his make-believe story before he ever got off a horse and set foot on the University of Nevada campus. What would best serve him, and the readers who enjoy him, is for Starrs to shed the pretense of being a scholar and just write what he writes best: pulpit polemics. And if this judgment seems too harsh and distasteful, and too much out of character with how he wants others to see him, then it behooves him to get serious about the most basic canons of scholarly research.

Academic Brahmins

> Even in the panoramas of the side shows and in folk ballads, the simple people, who are hardly simple at all, love the stories of the rise and fall of great men, of eternal change, of the cunning of the oppressed, of the potentialities of mankind. And they look for truth, for what is behind it all.
>
> —B. Brecht, quoted in F. Weinstein, *History and Theory After the Fall: An Essay in Interpretation*

ONE of the blue-ribbon banes of the social sciences is the immense amount of tripe that appears in major university press books and in what are alleged to be respectable academic journals. Behind, or rather beneath, all of this blather lie the highly paid producers, the chest-thumping impostors who not only receive disproportionate attention for their effluvia, but also occupy the high church chairs in the academy. Widely perceived as keepers of the Gates to Truth and Deeper Insights, they are, in their own minds, ever wise fortune tellers forever ready not only to reprimand but to direct and command the academy's untouchables as well. Protestations to the contrary, these self-anointed elites have little more than contempt for castes below them, social science brethren whom they scorn as theoretically impoverished empiricists, disciplinary pariahs worthy of little or no reward for their gritty hard work and foot-slogging data gathering that allow them to say new and often fairly interesting things about an ever changing, complex world.

These Academic Brahmins (ABs) follow a simple, invariant rule: Never get too close to data, facts, or the real world. Data and facts, however socially conditioned, are scary, meaning not only are they commonplace and therefore the stuff of ordinary people—the lower castes—but, worse, the factually rich worlds of the hoi polloi rarely bear much relationship to the

highfalutin "theorizing" that ABs confidently twirl and spin into their peculiar brand of fool's gold.

Academic Brahmin storytelling ("theory" construction) is inseparable
from language: the more obtuse and arcane the vocabulary, the more confused and convoluted the grammar—the latter especially—then the more
profound one's message and the less likely one is to be charged with theoretical pollution and the generation of erroneous factual claims. This
charge is particularly serious because it implies that one is slipping in caste
rank, perhaps even destined for that ordinary land of untouchables. Academic Brahmins aim to be not in the world but rather above it, structuring
what they can only imagine, or know second—or thirdhand. The world of
ABs is fashioned not in the common parlance of everyman's language, but
in a language that only ABs can really understand—or at least pretend to. It
is well understood among ABs that the facade, if not the fact of esoteric iso-
lation, decreases vulnerability, and so makes it easier to claim what cannot
be disproved by always vulnerable untouchables.

Thus, one almost universally distinctive trait of ABs is their special use
of language: invariably obscure, jargon laden, difficult to understand by even
the best native speaks of the language in which written, and chockablock
with code words and grammatical nuances that are acquired only through
hearing and reading the "right" texts. One sure test, then, of whether one is
a strict AB is the consistent ease with which facts, human complexity, and
the ordinary ways in which everyone else attempts discursively and de-
scriptively to portray the world around them are routinely ignored.

Academic Brahmins have no interest in showing or demonstrating that
this or that particular person is, as charged, blind or mistaken, or incapable
of seeing the world aright. It is enough to assert that it is so *in general,* and
beyond it, that there is a malaise in social science or in this or that discipline, one that is crippling to such a degree that no one of right mind can
possibly see the world or any of its components profoundly until the firm
directives of the ABs have been understood and strictly adhered to. Thus, in
keeping with their self-imposed first rule of not getting too close to data,
facts, or the real world, ABs eschew a listing of offending untouchables;
they direct their messages at a class, a clearly subordinate caste, one as
large or small as one might wish to imagine.

In the same way that untouchables are rarely identified by name for
their wayward ways of describing and conceptualizing the world, so it is
that specifically *what* untouchables must do to rise out of their lowly station in life, to overcome their never-ending predicament of not really "get-

ting the meaning" of the world they encounter, is never made explicit. It is never made explicit in anything remotely like an operational sense for the simple reason that to do so would require an example, or several of them, and it would have to come from the very world that belongs to the untouchables. And this world, as noted, is mundane, polluted, ordinary, and never important or "rightly conceived" in the way that ABs find elevated enough to embrace. The world cannot be rightly described or conceived by untouchables, or anyone else for that matter. ABs embrace not the ordinary world as ordinarily described in all of its quotidian messiness, but words, and more words—and words and grammars that by their very nature are elusive, rarely concrete.

Academic Brahmins like nothing so much as the mangled and thoroughly elusive wordplay constructions of other certifiable ABs: ones who are foreign (ideally French, or other European), have established AB reputations, and are best able to distance themselves consistently from the real world. Most assuredly, ABs do not want to come in contact with the untouchable world, for to do so would pollute their cozy and pure stories, which is another way of saying that they would be shown to be empty of content and meaning. To add content and meaning from the human world is to have to deal with immense diversity, and differences, which, by their very nature, militate against universalizing theory, the purity of the fool's gold. And here, then, to come full circle, is the very reason that ABs, in their off-limits inner sanctum, make so much of never getting too close to data, facts, and the real world.

Because the AB world is self-contained and quite apart from the real world—so permitting revision only within the limits of its own language—it is, like all worlds of belief (religion), not subject to refutation. In this regard, it is very unlike the world where untouchables live and work, where, in principle if not in fact, every claim to new fact, truth, or knowledge is subject to the discovery of yet another new fact or truth, any or all of which can significantly redefine the meaning of the evanescent world that preoccupies untouchables.

Academic Brahmins not only subscribe to fairly rigid norms about how to behave toward or engage the polluted world beneath them, but also hold tight to rules of engagement within their own small community. It is, for people who aspire to this priestly station, not merely a matter of knowing how to use the grammar and code words of the AB world that matters, but also knowing what to say about how one is behaving in or toward the world that they will not in fact, by design, engage. Thus, to a degree not found in the world of untouchables, ABs are hypersensitive about being politically correct, giving the most radical of feminists all of their due, and ensuring that minorities and ethnics of any stripe and every belief are—in word and word only—invariably treated with the utmost deference.

Purity in appearances is of paramount importance to ABs. They are at once haughty, elitist, and all-knowing to all of those disciplinary untouchables whom they are always berating and counseling, yet showily deferential to all others—the ones outside the academy who inhabit that gritty world of reality they do not know and do not want to get too close to. Academic Brahmins would have untouchables and everyone else believe that not only do they have a corner on human sensitivity, but their priestly sensitivity is so deeply felt that they "know" and can "feel" how marginalized people, oppressed women, and downtrodden Third World natives live their lives. They know it and feel it better than the very untouchables in their own disciplines who may genuinely be counted among the Other (that patronizing code word so lovingly embraced by ABs) or, if not, have spent the better part of their lives studying it.

I got to thinking about these Academic Brahmins and untouchables when I received an e-mail from a longtime friend. He wrote: "For something truly funny in a pathetic sense see the discussion of Soja's *Thirdspace* in the June '99 issue of the *Annals, AAG.*" I would soon discover that he was referring to the "Book Review Forum: On *Thirdspace: Journeys to Los Angeles and Other Real-and-Imagined Places,* by Edward J. Soja."[1]

My friend of many years is a full professor in a well-respected university and Department of Geography. He has written several books, all published by major university presses. Yet for all this accomplishment, and the score of academic articles that he has published in frontline academic journals in some thirty years of continuous productivity, he has no disciplinary reputation to speak of. A major reason for it is that he has never, to my knowledge, wanted to be party to AB prattle. Worse still, he's a hard-core "empiricist," and therefore a thoroughgoing untouchable. Nevertheless, he is one, I'd boldly venture, whose published books will long be read after AB sacred texts have become yellow and moldy from neglect.

I first met Ed Soja when I was a beginning graduate student. He had recently returned from Kenya and was writing his dissertation. I had just left a promising career in accounting and was taking my first graduate courses in the department in which both of us would get Ph.D.'s. I can still recall one day meeting Ed and asking him what his dissertation was about (modernization in Kenya).[2] He gave me vague and splashy answers that left me only confused, and for a brief time after our encounter I wondered if perhaps I was just too dense to understand anything he was talking about.

As the years went by, I rarely saw Ed, though from time to time I heard about his successes and failures. Now and again, I would come across something he had written, and what he had to say invariably left me with an empty feeling; he taught me little or nothing about the worlds around me or the ones beyond my intellectual and peripatetic reach. Before long, I began to sense that Ed Soja, though not quite sure where he was going, nevertheless didn't want to compete directly with someone such as David Harvey. Harvey had a good head start on just about everyone in geography on Marx, and it was clear to just about everyone that he wasn't going to be easy to catch.

Then one day in the late seventies at McGill University, where I was teaching, I had a lunch-on-the-lawn, under-the-trees chat with Ed. I learned that he was heading off in a new direction—sort of. He was still keen on being included among the liberal-radical crowd that had gathered around David Harvey, yet Soja needed his own turf. This McGill day, I think, was the first time that I really became aware of Soja's burning desire to find himself mired in obscure European intellectual gobbledygook. That sunny, warm McGill day he spoke with great animated enthusiasm about what he was going to do and what it all meant. When he got through, I concluded that he was lost.

Not too long after Soja's *Postmodern Geographies* appeared, I read it.[3] I found it just plain disgusting: bloated, poorly written, and, good untouchable that I am, irrelevant to anything I was doing or could imagine doing in my field-based studies. Nor could I see that there was anything in this book for any of the untouchables I knew, or whose work I respected.

For almost all of my academic life beyond undergraduate studies, I have applied a simple rule to what ABs, or anyone for that matter, have to say by way of instructing on how the world ought to be approached, looked at, or understood. That rule is: Be specific enough to tell me what I need to do to see the world differently, or aright. Or, lacking this ability in the form of a protocol or menu, show me by way of example, or point quite specifically to what someone has done that clearly and amply illustrates the approach being advocated. If I can get access to none of it, and I cannot make sense of what I am reading, or get anyone I respect to do so for me, then I conclude that I've been hustled and fed drivel.

I've long applied a variant of this rule to the claims of mathematical model builders, who, I've always reasoned, are no different from the rest of us with regard to their models. If a mathematical model builder claims to have discovered some new truth about the world, then he can state as much in ordinary language in his conclusion. If at this point I'm taken in by the novelty of the claim, but have doubts about the mathematical chain of reasoning and cannot work it out myself, then I'll seek expert opinion. If I am not given the "goods" in the conclusion, and cannot be convinced by someone using ordinary language that there are goods worth buying, then once again I conclude that I am dealing with a con man.

The hype in the *Book Review Forum* comes at us like a High Sierra spring flood. From one reviewer we get: "Few writers have had as great an impact on our conceptions of the spatiality of social life as Ed Soja. From his ground-breaking 1989 book, *Postmodern Geographies*, [Soja] has pushed geographers . . . to reappraise what we consider to be our traditional disciplinary objects of inquiry . . . ; [and] at the heart of Soja's work there lies a deep *moral concern* for human welfare."[4]

From another reviewer apparently trying to get into the hallowed corridors of the ABs, we're told that "Ed Soja's *Thirdspace* [is] a provocative addition to the tradition of innovative writing by established geographers." In addition, *"Thirdspace* is a multidimensional book that will doubtless provide fertile ground for an array of critical exchanges among geographers and other academics stimulated by the 'spatial turn' in the human sciences."[5]

From a third reviewer, also an AB (though seemingly at the door and perhaps ready to bolt), we're told that *Thirdspace* is "brilliant and inspiring" and "imaginative and inventive," and "contains a wealth of fascinating empirical and theoretical material." The book also "thrills and entertains and dazzles but ultimately tells us little, leaving us bewildered, overloaded with images, rhetoric, and vicarious insights, few of which seem to gel anywhere, least of all in our streets and cities."[6]

I don't know whether—with the qualified exception of Andy Merrifield whom I've just quoted—these reviewers can't read; consider themselves loyal, card-carrying ABs; or think of themselves as being on the threshold of formally being admitted into the chambers of this exalted caste.

You would think from some of these comments that Ed Soja, and Ed Soja alone, discovered that space (and its variants: region, place, location, landscape, and so on) not only is important, but also ought to be given its rightful place alongside history and society in the scheme of knowing how the world works. Somebody (including Soja) hasn't read Kant or the history of the discipline, or, it would seem, paid one iota of attention to the hundreds—nay, thousands—of empirical studies done by geography's untouchables just since Ed Soja was given his Ph.D. in the 1960s. All this "spatiality turn" that turns Soja and his admirers on is just so much old news, and to the extent that it's not, in that Soja and his less than completely reverent supplicants wish to confer on "space" some kingly throne heretofore allegedly unimagined, it is nothing more than the most insular and bloated kind of conceit imaginable. Geographers have been through all this spatial mania in Soja's academic lifetime (and mine), and all the smart ones who were subjected to this madness long ago came to their senses with regard to just how narrowing and parochial a hard "spatial turn" can be. They moved on to positions that put the meaning and significance of space in perspective, whatever was needed or required for the particular time, place, people, and problem of interest.

There is, contra Debra Dixon, one of the reviewers, no deep moral concern for the welfare of humans in *Thirdspace.* If Ed Soja would really like to get into deep moral issues concerning the welfare of humans, then he need only stay close to home, take off his academic pinstripes, and do some serious footwork in Watts or among Santa Monica's bums. Or he can always call me and come down Orange County way (my home), and I'll give him all kinds of names and entry points into deep moral and profound welfare issues concerning Santa Ana's Hispanics and its two hundred-plus gangs and the havoc they cause. (After all, marginality is a hard and very sensitive rub for overpaid, ultrasensitive ABs such as bell hooks [I know, she insists on being different with her little *b* and little *h*] and Ed Soja.) But for an AB such as Soja, doing so would be going too far; indeed, he's pressing his luck in saying as little as he did about Los Angeles and Amsterdam. To have said more, and said it more deeply and with greater descriptive originality, would have risked heresy, expulsion from the only caste where, apparently, he can imagine finding spiritual and intellectual comfort.

The reviewer whom Soja most wants to ignore—Andy Merrifield—and is most unhappy with had the following to say about Soja's squeaky foray into the land of untouchables: "When he does try to stoop that low and ground himself, everything sounds strikingly conventional (aside from the flowery language). Is this really just good old urban and regional description? Or is it alliteration masquerading as insight? I'm not sure."[7]

I am. And I think Soja is too, for about this criticism or any others that get anywhere near his entry into the real world, he is conspicuously silent. In *Thirdspace* (as he has on a couple of occasions in the past), he lost his way as a pedigreed AB; apparently, he can't completely shake off those mundane Kenyan days when he actually cared about the location of post offices and telephone poles and polluting First World pesticide cans.

Before returning to what Soja is disappointed about with regard to his reviewers' comments, it's worth saying a word about his "positioning," or rather how eager he is to signal where he sits among his reviewers and other ABs. We get this insight into Soja's elitist concerns when he just can't resist going on about one reviewer's comments on "thematic play on geographical Greek choruses and choreographics." Soja, in his own words, has only "one small complaint" to register, and it's that the reviewer doesn't understand that "the Greek root for choir and chorus is not the same as the root for place or territory, despite the tempting similarity in pronunciation."[8] Soja, in the previous line I've just quoted, gives us the Greek word for each.

I'd bet my right arm that Ed Soja knows about as much Greek as I know Polish, and on that score I know zilch (despite being through and through full of Polish genes and now and again hearing Polish spoken by my father and relatives). But even if it can be assumed that Soja is fluent in Greek and understands Greek root subtleties in all the ways I don't understand even fairly basic subtleties in Spanish (a second language I'm fluent in), I can't see that Soja's making this correction serves any meaningful purpose. It matters not at all to past, present, or future generations of field-hungry geographers or social scientists what the Greek root is for place, space, or territory. No, the only purpose served by Soja's showy Greek show is to demonstrate that he's full of affectation and lacks perspective on what does and does not matter.

Are these ABs half serious, or is it all an inside joke, a big put-on? From a reviewer by the name of Rob Shields, we get some genuinely stupendous claptrap:

> A rolling process, an organic entity, the spatial was too important to have been closed within reified notions. Even under scrutiny, lively and paradoxical processes kept a multipart dialectic in action as people, objects, knowledge and capital interacted at multiple spatial scales, flipped between figure and ground, and intersected with human life-projects to render intentionality, identity, and the finality of social structures always precarious.[9]

I don't sound this unintelligible even when I'm too drunk to stand.

But it's the very kind of wordplay that Soja just loves, and to which he responds by getting all wrapped up in the "rolling process," first getting into the question of whether *thirdspace,* the word, is meant to be "a reified noun rather than an action word"; and then whether it is a "recombinative or 'rolling' process of development and performativity." Then, not quite sure just how conservatively Brahmin he's been or should be, he's eager to apologize and return to the master of wordy spatial nonsense: Lefebvre, and all of his spiritual ruminations about love, poetry, and what-have-you stirred into Soja's cocktail that goes by the name of "Thirding-as-Othering" (a concept, no doubt, far beyond my intellectual reach of comprehension), which is, Soja confidently informs, "more psychoanalytically cryptic and intractable, more rooted in the microgeographics of intimacy, desire, sexuality, the body, the unconscious, yet less accessible to conventional geographical analysis." [10]

This kind of fatuous nonsense is patently transparent, which, to all of those aspiring and resident ABs, doesn't matter; nor does it matter to gullible untouchables eager to rise above their lowly, dirt-poor station in life or, barring this milestone, to at least get a chance to hobnob with high-priest Brahmins such as Ed Soja.

Another reviewer, Patricia Price, responds to what Soja has written not so much as a reviewer but as a high-horse screamer who's not much thrilled about Soja's "selective erasures" of . . . well, I guess, her feminist sensibilities.

> Yet for many feminists, both within and outside of Geography, the negotiation of pathways between the possibilities and the pitfalls of poststructuralism has been the subject of difficult choices. The stakes are very high for some of us: they involve our voices, our integrity, and our careers. Perhaps our generally wary (yet far from dismissive) stance toward poststructural approaches is born of decades of exclusion within modernist geography, and what many of us see as a dangerous trend toward reinscribing the Same at the center of poststructural geographies. [11]

I'd sure like to be told in a straightforward way exactly what's on her mind. If all this "high stakes" stuff has something to do with how being a feminist has affected her salary, or how her First Amendment rights have been abridged, or how she thinks universities are pretty oppressive places, then we might have a lot to talk about. But as I read her wooly prose, neatly wrapped in some packaging labeled "poststructural geographies," I barely had a clue what was on her mind. And that clue began to vanish when I started choking on the big *S* in her godlike "Same."

When I'd gotten the big *S* out of my throat, it occurred to me that if Patricia Price is indeed a feminist who cares about real people, then she might be willing to accept my invitation to visit Tijuana's Adelita Bar, the biggest bright-lights brothel between Mexico City and Nome, Alaska. There she could try out her "funhouse" (her word) poststructuralist geographies on the scores of young girls and women with children and without husbands and means of support who work ten hours a day, six days a week, dressing and undressing and hopping on and off beds in ten-dollar rooms with complete strangers to be able to buy food and clothing and pay the rent.

It seems that Soja, the apologetic, "gentle" interpreter of Lefebvre, has gone "beyond" his hero, a feat accomplished by two decades of hard thinking on "rebalanc[ing] our understanding of spatiality, sociality, and historicality," not by showing us how to do this "rebalancing" act, but rather by getting on a three-wheel wagon of "transdisciplinary perspectives [that include] psychoanalysis, literary criticism, discourse analysis, structuralism, existentialism, linguistics, critical historiography, social theory Marxism, and critical philosophy." [12]

This view into and across heaven and earth is, of course, completely preposterous. It is akin to claiming to be able to jump across the Grand Canyon, walk through cement walls, or speak with genuine fluency in twenty different languages. This kind of I-know-what's-going-on-just-about-everywhere is the most obvious kind of fraudulent intellectual crowing imaginable; it's academic machismo without a penis.

Equally suspect—to take another example of dozens that could be cited about the cockeyed storytelling that excites Soja and far too many people who ought to know better—is Soja's charge that geography "remains a masculinist, racist, and elitist discipline." To the extent that this statement is true (and a serious charge it is), then we should be given empirical evidence. But we get none whatsoever. To boot, the charge is an odd one in that in the very next sentence, Soja cries that "Critical dialogue is too rare—in Geography especially—to be reduced to mutual name-calling, no matter how entertaining such battles over Men versus Women in Space may be to the average reader." [13] Then why—other than to be politically correct and to ingratiate himself to other knee-jerk liberals who as often as not feel good about charging white hetero males (in particular) with racism, paternalism, elitism, and a whole lot more—would Soja charge an entire discipline with being racist, masculinist, and elitist?

In the end, Soja, for all he has to say about space and geography, has neither love for nor interest in geography. The discipline, for him, is simply a pounding board, a motley crew of sullied untouchables who are so lame-brain that they leave all serious discussion of space, place, location, and region to outsiders. And because it is outsiders who are in the know, Soja informs in the last line of his rebuttal, "geography has become too important to be left exclusively to geographers." [14]

My position is quite different. Geography and its mission as an empirical as opposed to a masturbatory exercise in storytelling that masquerades as social science profundity are too important, and have too much to do, to be able to afford Academic Brahmins such as Ed Soja.

I know several untouchables who proclaim that they will not waste their time reading the drivel produced by ABs, or, as I am inclined to do, war with them. This kind of position is a mistake, and a serious one. It's a mistake because ABs love to reproduce themselves, lest they have no one to talk to, no one to niggle endlessly with over whether Marx or Foucault or Lefebvre or Soja or Deleuze or hooks or Rose or whomever got this or that irrelevant piece of wordplay and confused grammar just right. And for every academic position these ABs take, it means one less position for someone eager and willing to find out something about the real world, and do so without a mouth and a mind full of moonshine.

It's important to engage frontally these intellectual impostors because they impart little of value to students or anyone else (other than how to be pretentious), and because there are plenty of intelligent people among the masses who see this AB blather for what it is and conclude that it is yet one more reason for denying support of any sort to universities.

It is also important to confront ABs aggressively, expose them for the intellectual pretenders and frauds they surely are, because one of the great missions of universities everywhere is not just to promote the search for "truth" and understand the world as best we can, but to do so honestly as well. And most of what ABs write and chatter about is just plain dishonest. But, alas, they're all too blinded by their immense egos, their elitist mannerisms, their twisted agendas, their still-born intellectual histories, and their self-serving interests to care one whit about what they're doing to students and the academy.

In the last analysis, blame lies not so much with these self-appointed

Brahmins as with all whom I have called untouchables, for almost to a person they're afraid to openly and aggressively make the case for what should be obvious to all. Like peasants with nothing but a hoe and a shirt and a bean patch, untouchables are convinced that their stake in the game is too small to matter and that it's just too risky to push their small pile of beans into the middle of the table and call the transparent bluff.

By My Sins Shall I Be Redeemed

Misereatur tru omnipotens Deus, et dimissis peccatis tuis, perducat te ad vitam aeternam.

May Almighty God have mercy on you, forgive you your sins, and bring you life everlasting.

—*Saint Joseph Sunday Missal* (A Christmas gift from my
    parents, 1955)

I was baptized a Catholic, I made my First Communion, and I got an Ad Altari Dei medal for putting in all kinds of hours on my knees as an altar boy for priests always eager to knock the water out of my hand because it would dilute all that good red wine I'd just poured into their gold chalices.

As a rebellious youth with ample carnal desires and a flair for adventure that bordered on the criminal and sometimes was, I held fast to the idea that it didn't matter how much I sinned because all I had to do was confess everything I'd done on Saturday, take communion on Sunday morning, and if I didn't steal yet another candy bar or Coke or give someone the finger before I died in a car crash later that day, then I was certain to make a direct ascent to Heaven.

After a while, and too many confession-sent Our Fathers and Hail Marys that just kept piling up in number, and the realization that on Sundays and maybe Mondays I was missing out on too much pleasure (that is, sinful acts), all this stuff about guilt became unbearable. Then there was the problem that I also started to take a real dislike to the priests and nuns who were continually trying so hard to make me feel guilty all the time. Like that time somewhere in my altar boy days when my mother and I went to a 5:30 weekday mass and there was no one to help the priest. So, thoughtful soul that I was, and eager to be in God's graces, and with a little push from my mother, I went up to the altar and served mass in my street clothes. Mass, that fateful day, was, I should note, for the benefit of the eight or nine God-loving Dominican nuns who filled the front-row pew on the right side, my mother and me, and a couple of old people who believed that seven days a week of mass meant that upon earthly death they'd find themselves several leagues closer to God than the ones who could manage to honor him only one day a week. Well, for that very serious infraction of church etiquette, I paid dearly. I was berated and made an example of in front of all of my classmates, and then mentally beaten up by my father on getting home because, Well, Rich, you should've known better. Where were your goddamn brains?

Many years later, I made my final break from the oppressive church. It was a break that came about not too many months after a six-minute marriage in a judge's chamber in Reno because my girlfriend was very pregnant. It came about when I went to the local parish priest and he told me he wouldn't baptize my newborn daughter because I was living in an awful state of sin because I had not been married in the church. At which point, and left with no reasonable alternative, I said, in quite polite terms, Shove the church. Within the hour—obviously unhinged by my brash rejection of God and church—the priest came to our apartment and interrupted my enjoyment of a beer, which now prompted me to tell him to get out of my house because I wanted to enjoy my alcohol and the company present, and that would be impossible with one of God's unforgiving messengers present. This contact with the Catholic Church and all its guilt-giving would prove to be my last, with the exception of all those times in later years in Latin America when I got curious about just how ghoulishly comical Jesus looked in one of his many glass-entombed impersonations.

I wouldn't know if there's a single academic who rejected the Catholic Church for reasons that resemble mine, and I don't much care or even find the question of much interest. But this fact aside, there sure does seem to be an inordinate number of academics in recent years who have willfully subjected themselves to some pretty heavy guilt trips by proudly wearing the label "colonialist" or one of its many fancy permutations. For the life of me, I've never quite figured out what was so special about the exploitative ways of Western colonialists toward "natives." I mean, these labels are merely context-specific terms for what all of us, everywhere, and in a very large number of ways, do to one another all the time—not just when we write, but also when we mistreat students and colleagues, when we turn greedy or hateful or revengeful, when we buy and then watch TVs that have been assembled by down-at-heel Mexicans in Juarez who are paid the grand sum of $4.50 a day for their gruntingly monotonous labor. All of these kinds of examples are every bit as much something to be guilty about (if you're into guilt) as anything you're likely to write about Kandyans in 1794, or at any other time in pre- or postcolonial Kandy.

The idea that it matters, or matters so greatly what one writes, and particularly compared to all the other things that we do daily in our lives to others, is just one more of those monumental academic conceits. It sure isn't anything to get all worked up about in the scheme of things, and to do so is nothing less than to flagrantly display a tremendous lack of perspective, both on one's quotidian behavior as well as on all those things that do matter and about which we might do something.

Jim Duncan, in a recent article titled "Complicity and Resistance in the Colonial Archive: Some Issues of Method and Theory in Historical Geography," is the latest in a long line of geographers, anthropologists, and other academics eager to publicly confess to a colonialist sin.[1] In Duncan's case, by doing some archival research on colonial Kandy, he's come to recognize his sinful ways, and he would now like forgiveness from some unidentified Almighty.

Duncan's not just asking for forgiveness for being "complicit with the colonial project" in some simple, straightforward way; he wants us to know that he's made some earth-shattering discoveries in his contemplative soul-searching. After a good thirty years or so of enjoying the many fruits of the easy life among geography's theoreticians, he's now ready to confess that he'd made an awful mistake, a particular kind of well-defined colonialist sin. He has learned that dirty fact finders (and I have long been one of them) who have always enjoyed the complexity of local geographies finely imagined and imaginatively narrated—but bereft of what he would call "theory" (I have only the vaguest idea what he has ever meant in this regard)—are now suddenly, shall we say, the academic archangels of this earthly enterprise called academia. Theory is out. Nuanced and detailed local studies are in. Duncan cannot, of course, come right out and say: Father, forgive me, for I have sinned. For half my mortal life, Father, I have been a shameless model of intellectual onanism. No, doing so would be far too direct a way of saying that in his overtly ambitious desire to be theoretically pure, and well recognized as a legitimate star among the theoretical cognoscenti, he was guilty of creating an endless string of cardboard characterizations and, alas, "objectifications," akin to making human behavior the simple product of half a dozen genes and almost nothing more.

Welcome back to reality, I guess I should say. I do fear, though, that if

Duncan is even half true to his claimed conversion, then he'll find that many of us who work hard at any sort of detailed take on this marvelous complexity all around are a rather grubby and even guiltless lot for the ways we go about getting information. Though we don't mean to offend people we talk to, and we try hard not to, we do step on people's sensibilities. And they do sometimes let us know that with our questions, we have gone where we do not belong. To some of them, and to the Jim Duncans, we are not only academic colonialists, but also unconscionable, predatory academic colonialists if we then publish what we've appropriated when we subject these unwilling natives to intrusive, offensive questions.

As in all confessions by people eager to be born again, there's a lot of Duncan's confessing that sounds more than a little creaky, disingenuous, as empty as the very "theories" in which he has invested just about the whole of his academic career.

The last two lines of his mea culpa read as follows: "We can see how overly ambitious theorizing can serve to obscure and bury even deeper indigenous knowledges and practices. It can allow a false sense of historical understanding."[2] Theory does not have to be "overly ambitious" to "obscure and bury." Just about any kind of theory that has been called such in geography, and the social sciences, obscures—period. And it obscures a lot more than just "deeper indigenous knowledges and practices." Furthermore, it is disingenuous to tiptoe with "It can allow . . ." Better and more cleanly put, Duncan should confess: Theory on the whole produces lousy history.

Duncan likes the company of someone named Barnett, who, in Duncan's words, claims that "the incomplete nature of the archives tends to raise fragments of information to a symbolic status." This statement is just plain wrong. It's not the incomplete nature of the archives that's the problem. It's the person working with the archives that's the problem. *Everything* we can possibly get our hands on in the present or the past, no matter how hard we try to get "the complete, complex picture," will be very incomplete. There is no such thing as getting complete access to any reality. "Symbolic status" will be an issue only to people like Duncan who have insisted in talking in such terms. It is not a problem to the ones who appreciate that, colonialist or not, every voice and every description is only as much or as little as we sensibly allow ourselves to make of it in light of all that we know.

Duncan says, "I also worry about academism, an orientation exclusively toward academic theory. This produces commentaries on commentaries on the world, often in rather abstracted language and with overly ambitious and decontextualized claims."[3] This explanation is a pretty good description of this very article by Duncan, to say nothing of all those other "theoretical" pieces that clutter his curriculum vitae.

He says, "To the extent that theory is in fact Western, one can say that the empirical is 'colonized.' "[4] Yes, and it is nothing to apologize for, either. I could just as easily, and logically, say: "To the extent that the theory is in fact Eastern (or non-Western), one can say that the empirical is 'colonized.' "

We are all fairly ugly beasts, and sinners to the bone, and not one whit less so because there are others we can patronizingly call the "Other," or the "native," or the Oriental. Today, we in the West make greater use of this weapon called language. Tomorrow it is the Other and the native who will use the weapon of language against everyone. And when that tomorrow comes, he will be no less kind and benign toward us, or any other native, than we have been toward him, as we are already discovering.[5] But then, beasts and sinners that we all are, we simply do the best we can—if we try hard to be honest. There is no one universal best by anyone.

If anything is "theoretical" in the way it is written, and with high-minded pretensions running all through it, it is this whole confessional Jim Duncan piece. And like all theoretical pieces, whether confessional or not, it contains more than a sprinkling of banalities. Here's one of several jewels: "Ordinary individuals within a colonized population may be, in fact are, obliged to say certain types of things, in certain ways, on certain occasions."[6] This assertion is obvious to the core.

Duncan is not forthcoming, penitent though he may see himself. He says: "Perhaps it is my contrary nature that I spent the first half of my career arguing against the under-theorization of cultural and historical geography and I appear to be spending the second half arguing against its over-theorization."[7] No, it was not Duncan's "contrary nature" that led him down the misbegotten path. It was, rather, a great career move consciously taken. It didn't require much intelligence for any of us to have seen fairly early in our academic lives that taking the theory-chat path—no matter how utterly vacuous—was just about the only sure way to the land of big kudos and high-profile recognition. If Duncan is true to his confessing words and flees the theory tip and tries to produce unpretentious prose, he will soon be just an ordinary academic native who will get colonized by "theory." He'll also be fighting like the rest of us to make his narratives a wee bit better and more convincing than fast-stepping journalists facing a nine o'clock deadline.

My betting money says that this article by Duncan is little more than another sin-filled and self-conscious career move. He and the people he reads, and who talk and think the same kind of gibberish, I'd bet, smell that it's now time to get theoretical about being nontheoretical. And the sooner one does so, the better, if one wants another pin-striped medal from the ones who claim that elitist colonizing house of theoretical claptrap that Duncan has always lived in.

When Logic and Language Go on a Walkabout

One should always cultivate one's prejudices.

—W. Somerset Maugham, *A Writer's Notebook*

EVERY discipline needs a good example or two of tortuous reasoning and bad writing and simpleminded thinking to share with graduate students, if only for the purpose of instructing them on how not to reason and write and traffic in insipidness. After all, knowing what not to do is every bit as important as knowing what to do, and it's far from clear that most academics know the difference.

To be most effective, it helps to use as an example someone who is high profile, is at a major university, and has had his work published by a first-rate press. A book works better than an article because more time and presumably thought has been put into the effort, there's more to chew on if the class has some slow learners, and it's a lot harder to plead that the author had an off day.

As good a candidate as any for a really bad book by a high-profile geographer who is at the very top of the professorial rank (Clarence J. Glacken and Bascom Professor of Geography) in a major geography department (the University of Wisconsin at Madison) is Robert David Sack's *Homo Geographicus*, published by Johns Hopkins University Press.[1] At 256 pages without notes and other paraphernalia, I'd guess the text comes to about 135,000 to 140,000 words, a major effort by most standards and one that at this length press editors want to trim unless someone has really convinced them that the message is pretty important.

Anyone with a decent sense of the history of geography in the past thirty years or so might guess where Sack is coming from and what the book won't be about just by taking note of the dedication: "To Yi-Fu Tuan, The Consummate Geographer." *Consummate*, the dictionary tells me, refers to someone who is "complete, perfect, or supremely skilled." Although there may be a couple of geographers who would be candidates for this lofty designation, most assuredly one of them would not be Yi-Fu Tuan. About the only real sense in which Tuan can be said to be "supremely skilled" is in the sheer quantity of words he has published, in his ability to skip around the edges of soft philosophy and perhaps mine libraries for esoteric sources.

Although he started in the profession as a field geomorphologist, it is clear that over the past four decades Yi-Fu Tuan has never been near fieldwork in human geography. So to refer to him as the Consummate Geographer can mean only that (1) Sack doesn't know what the word *consummate* means; (2) he knows what it means and believes that having significant field-based research in one's portfolio is a nice little extra but beside the point; or (3) what Sack sees and likes in Tuan he sees and likes in himself, and, of course, who doesn't want to see himself as the Consummate Anything? (There is, after all, a sack full of ways of complimenting ourselves in print.) I'll give Sack the benefit of the doubt on (1) (though I shouldn't, given how poorly he writes), and opt instead for both (2) and (3).[2]

There is, in all of this writing, another sociological subtext, that of elevating other geographers just like Tuan and Sack—Nick Entrikin and Michael Curry, the first a Sack student, the second a Tuan student—into that lofty realm where they too might someday be described as "Consummate Geographers," despite having produced no more than soft and irrelevant "philosophy" and also not having a clue what fieldwork is about or how to gather original data and why it's all so important to an empirical social science such as geography.[3]

In a book of this length, we could reasonably expect to come away with *something* new or genuinely suggestive about *Homo Geographicus.* But, in fact, what we get is one nearly interminable parade of banalities, throwaways, incoherent assertions, and muddled claims that add up to nothing at all. Sack's got this thing called "place" on his mind, and to judge by what he's got her doing, she's got to be one of the more delectable intellectual mistresses of all time.

Place, for Bob Sack, is a "force," an "agent," it is "destructive," it "constrains," it "enables," it "creates differences," it is something that "surrounds us," it is something that is "primary," it is something that is "secondary"; place is just about anything and everything as thing and process you might imagine. As someone who has an occasional interest in people and in places, I always thought that it is *people* who are "forces," who are "agents," who are "destructive," who "constrain," who "enable," who "create differences," who "surround us," and so on.

Sack's mind-numbing use of place as I have noted brings forth not a single insight of any sort about human behavior or places where humans live and procreate and die, or do anything else that's interesting. So why, then, does Sack use the word *place* in all those ways that suggest that he just doesn't know what the word means? (According to the *Random House Dictionary of the English Language, Unabridged* 2d ed., place is (1) a particular portion of space, whether of definite or indefinite extent; (2) space in general: time and place; or (3) the specific portion of space normally occupied by anything.) The principal reason, I think, is that Bob Sack is just plain confused. I have singled out his use of the word *place*, but in fact he's muddleheaded about all kinds of things, and rather consistently so. And it make me wonder whether it's possible that Sack, the self-anointed philosopher of geography, ever once read Bertrand Russell, who said that clear reasoning is one of two great virtues, or whether he in fact might have something to say but just never learned how to become semiliterate in his native language.

Like Tuan, Entrikin, and Curry, Sack somehow just can't deal with real people, and so he goes running off in every direction to find something airy, pretentious, and meaningless to put in their place. He apparently can't even appreciate that the title of his book is *Homo Geographicus*, and that although there are all kinds of places where man is not now found, places are, with regard to issues that he very much does want to discuss, interesting, odd, or different because of *people*. And because of what people do, not—as Sack would insistently have it everywhere in this book—what places do. Places, to put it bluntly and without qualification, don't do a thing.

With no knowledge of the real world, nor apparently much real interest in it, Sack is left to dig continually in infertile soil, and say the nuttiest things. Here are some examples:

"Genes form the basis of biological life."[4] Is there another kind of life beside "biological" life?

In the very next sentence, we are informed that "[Genes] are obviously located in space." Could they be located in nonspace? Oops, I forgot to add that after the word *space* there was a comma, followed by: "from the level of the cell to that of the organism and its relationship to others of its species." What could it possibly mean to say that "genes . . . from the level of the cell"? This sentence is just plain nonsense, and wrong.

But Sack doesn't let up (and neither will I, yet), for the next sentence reads: "In this sense genes possess a geography, even though the concept of a place does not obtain in the nonhuman realm because there is no real territorial structure with rules of in/out of place." *What* could this sentence possibly mean? I don't know, and I teach biology to university students.

Next comes: "But the genetic material of humans is organized by place as well as space." Does he mean to say that genes or alleles do show up in fairly predictable ways on DNA?

These kinds of platitudes and inaccuracies masquerading as much ado about place continue, and I cannot help but wonder why any decently sane editor—or the many people Sack gives credit to for having read "various drafts," including fairly prominent geographers John Agnew, Nick Entrikin, Yi-Fu Tuan, and Karl Zimmerer—didn't, upon reading this and many other paragraphs just like it (and the book is full of such pearls), say: Bob, cut all the bullshit.

Cut, cut, cut . . . and there's still more to cut.

"Ever since the first hominid, genes have not simply been carried by individuals of the species through space, but have been localized in primary places, if only in settlement sites and hunting areas designed and controlled by families or clans." Well, you bet, Bob. If I decide to go on a trip, I assure you I don't have the option of getting inside my DNA and taking out a few genes and leaving them behind on my desk.

"These simple places, or territories, serve to select genetic material for breeding among and within human groups." No, Bob, sorry, but places or territories do not select genetic material; people do.

Then, finally, we come to the last sentence in this one paragraph: "More recently, as a necessary part of the rise of agriculture, humans have created places such as fields and gardens that arrange and alter the gene pools of plants and animals."[5] Once again, we are faced with the magic power of places, which, in the form of fields and gardens and places, arrange and alter the gene pools of plants and animals. All this information, I venture with great assurance, is going to come as quite a surprise to geneticists I know, and to my wife who works in both the field and the laboratory on the genetics and behavior of finches, and also to Sack's colleague Karl Zimmerer who has more than passing interest in how Andean peasants manipulate the genetic makeup of their subsistence crops.[6]

Sack is, alas, irrepressible, and he doesn't let up with his triteness. Four paragraphs later, he opens with this jewel: "An important characteristic of a place of solitary confinement is that the prisoner is there involuntarily."

Really? I thought that when we use the words *solitary confinement,* we're referring to places to which people retreat voluntarily to write books like *Homo Geographicus.*

Thumbing through the book at random, there is no shortage of Bob Sack originals.

"Places affect each other because they are connected in physical space."[7] No. If two places are ten miles rather than ten thousand miles apart, then people in those places that are near one another are likely to interact more than people in those distant places. It is *people* who are affecting one another, *not* places.

"Places of the earth possess a surface, an appearance."[8] I thought all along that places were just black holes, a blank before the eyes and the mind.

"What do places mix together? Virtually anything."[9] No, again. People mix, and when they do so, it might be over just about anything, especially sex in which the "agent" is, I assure you, not place. Chicago and Los Angeles don't mix a damned thing, but *people* carry the HIV from one place to another place.

"The objects and events of the universe and the world have location and extension in this space. Trees, grass, mountains, and soil occur in it and extend through it."[10] Gee-whiz, I thought that I could sleep in the exact place where my computer and printer lie because, after all, my computer and printer are dimensionless or, shall I say, occupy nonspace.

Well into the book, Sack finds himself with his daughter at home on a Saturday afternoon, and she has a homework assignment to write an essay about "home." Because Dad Sack runs on for nearly five thousand words about his daughter and her "home" homework assignment, in a book ostensibly all about place, you'd think that we'd get some genuine insight into the meaning of home. But no, nothing of the sort is forthcoming. Instead, we get just one rambling set of shopworn "philosophical" statements: about truth and linguistic convention and mathematics and Platonic ideal types and Kantian expanded thinking and justice and merit as a yardstick in schools, and his daughter's enthusiasm, but not a single Bob Sack insight into this kind of place called home.

I'd like to believe that Bob Sack wrote this incredible drivel called a book, a book that someone out of his or her mind at Johns Hopkins University Press saw fit to publish, because he needed the money, and, tongue-in-cheek, laughing all the way through the long exercise, he saw dancing images of places on Lake Mendota. Not where he could build his cabin, but rather where these places as "agents" with "force" and a lot more would build the cabin for him.

But, alas, there's plenty of published evidence to confirm that *Homo Geographicus* is the true Bob Sack in all of his tight-jawed and very muddled seriousness.[11] He's absolutely dead serious about the mighty message delivered, and—I'd bet—he was hopeful that the world would sit back and read what he wrote and say, Wow!

I did indeed sit back and, not knowing whether to laugh or howl, said, Make that another Bud.

Words in the World, Worthless Words

Deceivers are the most dangerous members of society. They trifle with the best affections of our nature, and violate the most sacred obligations.

—G. Crabbe

MY nine-year-old son, Cole, has some nine hundred Pokémon cards. He claims that their current market value is $1,820, and that a couple dozen of the cards, mostly the holographics, have values that range from $15 to $50 each. His knowledge of the cards (what's on them) and how they relate to one another is simply prodigious. Give him the name of a Pokémon, and he will provide all kinds of quantitative detail about it. By any measure, he has a crystal-clear memory, one of the many marvels of a young and very good mind.

A week ago, on a Saturday morning, I told Cole that if he wrote me a one-page short story about anything, then I'd give him $6 to buy a pack of ten Pokémon cards. He had the story written before lunch. The sentences were strong, the flow was excellent, he moved easily from the real to the imaginary, and what he wrote was very much a story.

Cactus Attack   ◎   Cole Symanski

Once I went butterfly hunting with my old friend Julian. It was a cloudy cool day. We went to a bunch of hills. It was a great place to catch some insects. We started out by catching moths and beetles. The moths were gray and brown and had one circle on each wing. The beetles were ordinary, except one. It was a rarely seen red click bug.

After searching for about 45 more minutes we came across a pretty big grasshopper. We chased it for about 5 minutes. It came towards Julian. Julian was 2 inches from catching it but it hopped away. And then I saw a Swallowtail. I was after it in a flash. Crash! I hit a cactus just as I caught it in my net. It was the smallest Swallowtail I had ever seen. The cactus hurt so much I saw Little Red Riding Hood circling above my head like a butterfly. Then you won't believe what happened! I went behind the cactus that I crashed into and I saw some people filming a movie of Little Red Riding Hood. For helping with the movie, we got one million dollars and a pack of Pokémon cards.

Five nights later, we went out to dinner, and as Cole engorged himself with more than half a pound of the best smoked salmon between San Diego and Malibu, I asked my wife several questions about the finch book she's writing. Somewhere in the middle of this conversation, Cole pushed aside the salmon he was devouring and said that he too was going to write a book. Great, I said, and then before I'd given it much thought, I added, Write a book of 150 pages and I'll give you $1,000. His handsome blue eyes grew large, like I'd finally bought the long-desired tickets to Cabo San Lucas to go after blue marlin.

Over the course of the next couple of hours, Cole asked me several pointed questions about what I had said. Was I *really* serious? (We tease each other incessantly.) Did he have to write a novel, or could it be a collection of stories? (One story, twenty stories, a hundred stories, I told him, and about anything.) How could he spend the money? (On anything, I told him—except candy.)

Cole apparently thought about all these promises that night when he went to bed, because the first thing he wanted to know the next morning was whether I was in fact really serious about giving him this huge amount of money for writing a mere 150-page book. And could he use all of it to buy Pokémon cards? Of course, I assured him.

After he went to school, to prove my honest intentions, that it wasn't just another running joke between us (of which there are now more than I can count), I wrote up the agreement and added in the tantalizing tidbit that the day he gave me the 150 pages, he'd find 1,000 brand-new $1 bills on the dining room table. I printed the agreement on gaudy blue paper and put it next to his computer, where he'd find it when he came home from school.

I have my doubts that any of it will come to much, though my wife thinks that Cole will in fact write the book. We both agree that even if he writes only, say, seventy or eighty pages and I wind up giving him several hundred dollars (which I will if he writes this much), it will be one of the best investments we've made in his education. That is how much both of us value writing, and particularly good writing.

I'm a Johnny-come-lately to an appreciation of really good writing. In fact, I'm not ashamed to admit that Cole at nine writes better and stronger sentences than I did at thirty. I can barely imagine how far ahead of me he'll be by the time he's half my present age. Late bloomer or not, writing is something I work at daily, and in one sense I am forced to do so. Although my advanced degrees are in geography, I have not taught a formal geography course for more than two decades. And though I am in a biology department and do teach courses in ecology, my principal university instructional duties involve teaching writing to senior biology majors.

Among the many little things I try to impress on students is the need to write clearly, forcefully, and with as much specificity as possible. I admire sparse writing, an aim often elusive in my own writing. I constantly remind students that if they're looking for God on a written page, then they'll most likely find him living among the details. I also stress that good writing, among other things, is like a well-toned athlete: lots of muscle and very little fat. Flabby writing is unattractive, and rarely seductive, I tell them.

In 1996, Michael Curry wrote a book with the title *The Work in the World*. He opens the second paragraph of the preface by stating: "My subject here is the nature of the written work in geography. . ."[1]

It so happens that long before I learned how to write some fairly decent sentences, I too had an interest in the written word in geography. In the 1970s, in a paper that caused dyspepsia and tribal shivering in more than a few geographers, I took Meinig to task for his writing, claiming that it's chockablock with hyperbole. Then I could not have predicted the postmodern turn and with it all this interest in texts, writing, and the instability of language. But now that we're all swamped and preoccupied by these concerns, I too cannot ignore them.[2] Nor, ostensibly, can Michael Curry.

His book has something to say about how people saw the stable written word before the postmodernism enlightenment, how all of it changed into a rather fluid and uncertain wash of relativism and too many voices, and why, briefly, it's not a particularly good idea to buy into an agenda of radical relativism. Too, he reminds us (some of this stuff is pretty banal) that writing occurs somewhere rather than nowhere, that writing "talks" to some people and not to others (the old community issue), that the same writer can write in different voices even in the same piece, and that any piece of writing in a discipline such as geography is situated within a larger world of fussy journal editors, moneygrubbing publishers, the historical moment, and so on.

But what do we learn about: How geographers write? How what geographers write by virtue of writing geography is different from what sociologists, anthropologists, or journalists write? How, for example, the hot boogie-woogie writing of Paul Starrs, editor of the *Geographical Review*, differs from the cool and sparse writing of his mentor, Jim Parsons, and what these stylistic differences substantively mean to a reader? What effect the imprecise and wooden writing of Bob Sack has on his messages and our willingness to plow through the rocky turbidity? Why some of the excellent writing we have seen in students who have gone to Berkeley may be a disguise for a lack of analytic acuity?[3] What is distinctive about the elusive and meandering writings of Yi-Fu Tuan, who was Curry's mentor? What, with examples, constitutes postmodern writing as opposed to modernist writing, if there is such a difference, and if there is what does it really mean? What, with examples, constitutes good, very good, bad, and very bad writing in geography, and why?

About any or all of these issues, Curry says nary a word. And so, *The Work in the World*, because it goes nowhere interesting when not addressing a single one of these questions (instead giving us a lot of stale, philosophical meal about place and space), I was left feeling as cold as I have felt when fishing for king salmon in winter off the northern California coast.

I suspect one reason for the total absence of what I hoped to find in *The Work in the World* is that Michael Curry has a hard time knowing where to find God on the written page. His writing on the whole is pretty vague and loose. More than once while reading the book, and other articles he has written, I've gotten a mental image of Michael Curry, whom I have never met nor seen a picture of, as someone who can't recognize an exercise bike when he's sitting on one.

In two places in the book, Curry does mention me, once for what I wrote about D. W. Meinig and once for instructing the opportunistic postmodernist Michael Dear to learn how to get some facts and find his way out of jabbering town-hall meetings.[4] When I saw my name on the page, I was hoping that Curry might show me how now and again I'm as blind about writing as my right eye daily confirms I am about seeing a full screen. But, alas, there was absolutely nothing of substance, leaving me with the empty feeling that other than to add a couple of more lines of text and a few more references to the note section, there was no reason for Curry to have even mentioned me. His analysis here is as shallow as it is when he briefly discusses the work of that modern-day love-in, hug-and-kisses-for-all pragmatist Richard Rorty, who reduces truth to any old kind of consensus—whatever feels good at the moment among the sitters around a campfire.[5] Curry's analysis, unfortunately, doesn't get one bit better when it comes to any kind of careful distinction between the fairly stable if slowing changing truths found in the natural sciences and those prejudices parading as profundities so widespread in the social sciences. Like Rorty, this rollicking sweep on the postmodern seascape is not a conceptual wave that Curry wants to or knows how to surf.[6]

So the book, as an inquiry or exploration, or even as a set of Michael Curry one-sided opinions on writing in geography (which I'd love to hear, no matter how much I might disagree), is a disaster. And the book would have been, for me, a disaster without qualification were it not for some God-sent details that appear in his lengthy chapter on Henri Lefebvre. Deep into this chapter, Curry discusses the kinds of references found in Lefebvre's most holy text, *The Production of Space*.[7] In particular, Curry has the following to say with regard to Lefebvre's rambling, disjointed four hundred-page take on space:

> The impression that we are given [by Lefebvre] is that (in 1974) there had been no serious thought about space, that those who have thought about the matter have failed to grasp the issue—in part, of course, because they have not aligned themselves properly with the greats. Indeed, this view of current works is heightened by the cursory attention that he pays to these authors; rarely does one get the impression that he thinks it worthwhile to devote time to their analyses.
>
> There is, actually, another side of the matter, and it is Lefebvre's silence about a substantial number of works available at that time that in various ways did indeed grapple with the issues that he was considering. He was conspicuously silent about James Blaut, William Bunge, Anne Buttimer, Walter Christaller, Richard Hartshorne, David Harvey, J. B. Jackson, Edward Relph, Yi-Fu Tuan, and Edward Ullman, all of whom had written (sometimes substantial) widely available works. Both of these practices, his negative references and his silences, are additional means by which Lefebvre situates himself in a social space, in this case a social space where he is the only person doing serious work *on* space.[8]

In other words, Lefebvre is one lousy and dishonest scholar.

In my undergraduate writing classes, Lefebvre might well get an A for posturing, but an F for ratio of fat to substance, and for scholarship. My saying so wouldn't matter all that much if Lefebvre hadn't, in recent years, been elevated to that ambrosial firmament where Heaven-sent mortals forever rain down indisputable truths on present and all future generations. And, alas, that is exactly what has happened, and it has happened because geography's current generation of self-appointed high priests and priestesses have made so much of Lefebvre and this book, *The Production of Space.*

So we have the likes of Neil Smith, Ed Soja, Don Mitchell, Derek Gregory, David Harvey, and several others whose names don't readily come to mind getting up on park benches and before starstruck gatherings at conference meetings—no notes in hand—proclaiming that Henri Lefebvre, that great French scholar and intellectual, not only is to be taken seriously but also is in a class by himself. Or as Curry reminds us, it is David Harvey, Mr. Geographer on Marx and Matters Related Thereto, who wrote an afterword to Lefebvre's book, calling it "magisterial," providing a biography, and then securing Lefebvre's place in academic heaven by concluding with a list of sixty-six books written by Saint Henri.

It would have been awfully nice for David Harvey to have gone to his pocket dictionary and found another word for *magisterial* in light of Lefebvre's intellectual dishonesty. But, alas, it's much too much to expect from Harvey or any of these other high-profile geographers who have been so conspicuously slobbering over Lefebvre. What seems to distinguish the lot of them is that when it comes to the written word, and the spoken word, it's all about polemic, ends justifying means, godawful capitalist exploitation, canonizing yet another woolly-minded European pseudointellectual, subordinating the messy empirical world to oleaginous abstractions, and making much sanctimonious ado about very little indeed. Scholarship, remembering the past, giving credit where credit is due, fairly representing the works of others: all are just about the last things on these Marxist agendas I keep bumping into.

Ruminations on a Misshapen America

Father Berrigan gives apt and timely warning to us of the danger of being "turned into a prestigious figure who cannot be seriously questioned, least of all by those bold and impertinent youths we seem to be producing of late in America!" He goes on to say: "I've seen [it] happen again and again—so blatantly and outrageously—on our campuses, where college professors simply stand on their records, their books and articles, their capital, so to speak, and rage at anyone who questions them in a searching, face-to-face manner. And so often, those professors feel called upon to vindicate not only what *they* have won (what they have come to!) but what others have won—the system!"

—G. D. Berreman, "'Bringing It All Back Home': Malaise in Anthropology," in *Reinventing Anthropology*, ed. D. Hymes

ANYONE who sets out to write a massive three-volume work that is expanded to four volumes after the first one is completed, each volume running to about 250,000 words, one of the largest single projects ever attempted by an academic geographer, has won the day against all but a tiny handful of critics even before the first lines are read. After all, who in their right mind can claim to find fault with someone who has dared to take on a task of this magnitude, in the case of D. W. Meinig and *The Shaping of America*, one that through three completed volumes has so far consumed some twenty mature years of his academic life?[1] The sustained and focused workload alone deserves only unqualified praise. And on this account I throw kudos at Meinig, as forcefully and generously as anyone.

The task of critiquing an effort of this description is further complicated if the critic is on record as someone who has taken issue with the author previously. And I, or rather that genetically identical soul mate of mine who goes by the name Richard Symanski, have the rather dubious dis-

tinction of having written that—to reduce the argument to its most elemental point—Meinig doesn't use language honestly; he's full of hyperbole
and other stylistic peculiarities that make his renditions of history suspect
at best.[2] But having now noted that I am hardly an unbiased critic of
Meinig, it's worth highlighting two caveats: No one is or can be objective in
any of these kinds of exercises, and arguments to the contrary are insup-
portable. Furthermore, what I have to say in what follows can easily be
checked against the work itself.

An impertinent question, I suppose, is: Who will carefully read most of
even one of these volumes on the shaping of America? Each one comes in at
about 225,000 to 250,000 words. Put simply, few people have the time to
read this much, not, at any rate, unless they are teaching a class on the historical geography of North America, claiming to be doing research on such
a sprawling topic, or retired and in love with reading and deeply interested
not only in American history but in one that is notably lacking in human
drama as well. Meinig's almost obsessive and unbalanced attention to what
might be called the "political and social geometry of American history,"
his poor understanding of how to effectively use and quote historical documents, and his utter lack of interest in the quotidian affairs of ordinary
people make these books "hard reads."

To boot, these books are written by someone who has none of the name
recognition of, say, an Arthur Schlesinger or a Paul Johnson, nor do the
books have the narrative fluidity and well-chosen arresting examples found
in, for example, David S. Landes's engaging best-seller, *The Wealth and
Poverty of Nations: Why Some Are So Rich and Some So Poor.*[3] People read
big, time-consuming books because of expectations about what an author
will deliver (based on past reputation), or because, in the case of Landes's
book, the stories he has to tell are very good reads.

Except as a kind of reference book to be sampled here and there, *The
Shaping of America* is a poor candidate for any kind of classroom use. It's
too big, too diffuse, and too hard to know what to do with its pieces. Anyway, I am certain, Meinig had no intention of these volumes being classified in any family of books related to textbooks. Whether they might be, or
in some sense should be, seen as quasi textbooks is another matter.

My guess is that Meinig very much saw this project as one that upon
completion would give him an almost unique place in the history of academic geography. Competition for efforts of this size (forget quality) is practically nonexistent, and it's very easy to imagine all kinds of people now

and in the future drawing favorable attention to the books, utterly irrespective of whether they have read any of them. I'd venture that in decades to come, not more than one out of ten practicing academic geographers (perhaps one in twenty, to be more realistic) who cites these volumes as "magisterial," "monumental," or "incomparable in accomplishment" will be able to discourse critically on them for as long as five minutes. They will no more really know what Meinig did or did not say than will the legions of self-important pretenders in the academy who claim to have read and understood *Ulysses* or *Á la récherche du temps perdu*.

But who will or will not read these volumes is irrelevant to Meinig and his own aims. If I'm right about what he wants out of the effort, then he'll certainly get it. And it'll be unqualified plaudits for the size of the undertaking, the uniqueness of the effort, and by default the significance, substantively speaking, of what he has done. The long-run rewards he will receive for this effort will have surprisingly little to do with demonstrated or demonstrable worth, whether *The Shaping of America* is, in fact, good history, good historical geography, and not only well executed but also as new and unique as Meinig wants to believe, and claims.

The Shaping of America is mediocre narrative history. Good narrative history is no different from good fiction, which has the quality of seamlessness, an uninterrupted dream. It is not just a matter of good and convincing writing, of forward momentum sustained, of tension and lulls appropriately placed and spaced; it is avoiding all of those easy-to-make missteps that create a break in the dream. All good historians know this rule, which isn't the same as saying they follow it religiously or that the ones who are really good don't know when they can get away with ignoring it. Meinig must know the rule; he just doesn't know how and when to break it.

Meinig continually and unnecessarily disrupts his narrative, sometimes in major ways, at other times in ways less unnerving. His principal problem in this regard, I suspect, is that his long-held desire to be respected as a rather traditional historian (historical geographer, he'd say, in deference to his training and academic history) has often been overridden by a desire to be seen as "theoretical" or "conceptual," contributing to a dialogue that supposedly raises the historical moment and all of its fascinating facts above the particular. And it is precisely here that he gets into trouble and writes narrative that is distracting, discontinuous, off-putting, and ultimately enough to have one continually put the book down, only to one day realize that interest has flagged, too much has been forgotten, and there's no good reason to finish the chore.

A good example of this kind of narrative disjunction occurs very early in *Volume I, Atlantic America, 1492–1800,* when Meinig makes much ado about what he calls seafaring, conquering, and planting, which he explains as follows:

> *Seafaring, conquering,* and *planting* [the last an ill-chosen word for its other connotations and what he's trying to convey] may be taken as convenient labels for three kinds of activities undertaken by various European societies overseas. They can also be taken as three phases in the European encroachment upon the American seaboard, for seafaring is the obvious first essential and conquest a necessary prelude to any extensive planting. And since planting of European settlers is clearly the most direct and effective means of rooting European culture onto the American seaboard, we may take seafaring, conquering, and planting as three essential components in the accomplishment of that end.[4]

This kind of "general formulation," which Meinig runs on and on about for more than a thousand words shortly after he opens with Columbus and then Cabot sailing forth, and returns to now and again in the text, adds virtually nothing to the story to be told. Indeed, the "concepts" just seem simpleminded, necessarily true, leading nowhere, providing no larger insight—and it is the larger insight, after all, that is the purpose of introducing or developing such "concepts," "formulations," "theory," call them what you will.

By the time Meinig is sixty-five pages into this first volume, and telling us about the Dutch and the Atlantic world of 1630, the reader is suddenly jarred out of the unfolding historical moment with a section titled: "Generalizations: Models of Interaction," which opens with the trite statement: "The Atlantic World was the scene of a vast interaction rather than merely the transfer of Europeans onto American shores."[5] Here we get the use of a simpleton's view of the world ("merely the transfer") appended to a "profundity" ("a vast interaction"), all by way of opening the door for Meinig to recall his "simple formulation of *seafaring, conquering,* and *planting,*" which he then further subdivides into "recurrent general patterns" of exploration, gathering, barter, plunder, outpost, imperial imposition, implantation, and imperial colony. With no sense of where it will lead, or how it will lead to new understanding (justify breaking the dream sequence, the good narrative), we then get treated to two full-page "theory diagrams" (my words): diagrams with Meinig's famous arrows, incomplete sentences, and portentous words such as *Prelude Phase, Fixation Phase*, and *Imperial Imposition*. All of it is far more confusing than enlightening. It's as if we had been pleasantly listening to Mozart when a young son charges into the room, yanks off our headset, and turns up the volume on his boom box. The screeching heavy-metal sounds make us want to not kick the kid out and return to Mozart but rather head for the kitchen and a strong drink.

We get several more of these kinds of breaks, one coming about halfway through volume 1 with a section titled: "A Geographical Transect of the Atlantic World," where, besides distracting text, we get "center-periphery concepts" outlined, and then a two-page spread with some forty or so boxes filled with words and arrows, a kind of outline that may well have helped Meinig in organizing his thoughts for writing but on the page of this book seems more like a kind of banal personal poetry than anything else.[6]

No doubt all of it means a great deal to Meinig, for various of these kinds of twittering diagrammatic models appeared in articles in high-profile journals prior to undertaking this multibook project. Someone must have convinced him that just because maps are often of value to geographers and others, then so too are all manner of silly pictorial representations—the fancier and more elusive, the better. Alas, all they really do is confuse and break the narrative flow, jar one of the dream of the good story he might have told.

One final example, this one from *Volume II, Continental America, 1800–1867*, concerns a section titled: "Geographic Patterns Within Imperial Territories." Meinig in this section is trying hard to make much ado about the word *imperial* and its various permutations, all by way of getting around to the fascinating story of how Indians were marginalized, subdued, and killed. Here, as elsewhere in one of these unnecessary and off-putting digressions, he writes as if he were in the classroom, as if, yes, he's really not quite sure if this effort is not a textbook after all.

> If we are to understand the geography of empires we must also focus on the most important features in the daily life of empire: the basic patterning of contacts and relations between the imperial people and the cap-

tive people. For convenience these may be viewed in terms of some commonly recognized categories.

Imperialism is, first of all, a *political* phenomenon, an unequal power relationship between two discrete peoples organized in territorial terms.[7]

Meinig goes on and on for another thousand words or more with this threadbare claptrap, with the aim of giving us "theory," until, finally, he wakes up and starts talking about the Jacksonian era and U.S. foreign policy that saw Indians as "resident foreign nations." *Now* Meinig is returning to the story, to substance—treating us as intelligent readers who can make up our own minds about whether we see U.S. policy, domestic or otherwise, as imperialistic.

Meinig now and again interrupts the dream of his story by, willy-nilly, pausing here and there to shift from the more serviceable third person to the familiar first-person plural, we. For example, in getting "theoretically" offtrack by talking about the "process of convergence of this imperial system," with a brief example of Genoese merchants and financiers to Lisbon and Seville, he suddenly writes: "[W]e can follow the shifts of Mediterranean seafarers to Atlantic bases." The next paragraph begins, "And we can trace the development of new tools." And then two paragraphs later, he begins, "We can see that region take shape in the turbulence."[8]

I can find no purpose to Meinig wanting to pull a reader into the narrative flow in this way; in fact, I would think many discerning readers would resent the "now do you see it?" didactic tone. *Just* tell the story, I want to say here and elsewhere when Meinig needlessly shifts to the first-person plural, or—less frequently—decides to put on his Anglophile hat and write about what "one finds" or "one sees" or "one understands." Done very sparingly, the shift to "we" works. But Meinig doesn't know when to do it. It's as if these kinds of sentences were written in the minutes before he was to give another lecture on the historical geography of the Americas to several hundred sleepy-eyed undergraduates who indeed need to be reminded that they're in the auditorium for a purpose.

In the acknowledgments of volume 2, Meinig is obviously pleased with a journalist who reviewed volume 1 and liked "the plethora of maps which broaden and sweeten the text."[9] Not only is Meinig far from unique as a geographer in believing that a lot of maps makes a study uniquely geographic and therefore different from history or other kinds of social science, but

Meinig, I think, is also trying to draw attention to something else, namely, the *kind* of maps that appear with such frequency in these books. They are maps that are chockablock with all manner of thin and thick and curved and word-filled arrows. Rather than enlighten and complement the text, however, the arrows beg, and insistently beg, for interpretation—and, even more so, for data. How, any serious student of history must ask, did Meinig decide to "value" the arrows—their thickness, their inclusion, their swirling, current-defying directions? Meinig provides not a single clue. This history is bad, and Meinig's medium in this particular instance for doing bad history is one of the geographer's key tools: cartography.

On many of the maps, there are so many arrows and they are running in so many different directions that it is awfully hard to look at them and not get dizzy, and feel awfully confused. Did Meinig ever once ask a trusted confidant or all of the people he so generously thanks in his acknowledgments if they could really understand what all these "shooting arrow" maps mean, and whether they provide additional insight (as with all good theory)? I am convinced that, collectively, these shooting-arrow maps by Meinig will someday be judged as one of the most embarrassing cartographic displays ever published by a geographer.

One reason for Meinig's lifelong love affair with his shooting-arrow maps is that as early as the 1960s he was preaching to graduate students that these kinds of maps show "process," whereas *mere* distributional maps merely show pattern. In Meinig's worldview, it all comes down to equating process with explanation and pattern with description, and what academic in his right mind is not going to come down on the side of process, seeing it as inherently more important than description? Meinig has never appreciated—and certainly does not in *The Shaping of America*—that serious attention to detail, and a lot of it well woven together, in fact constitutes explanation. Description, after all, is just process "stopped," a momentary frame for viewing. If description is well done, and sufficiently dense, it is, in the realm of history and the social sciences, just about the best kind of explanation imaginable. But Meinig doesn't get this fact, and one good piece of evidence is the plethora of all these "shooting-arrow maps." I guess he believes that his critics (admirers on other matters) will easily embrace his map ideas about pattern and process, description and explanation, and therefore give him credit where credit is not due. And on this score, given how stone-headed most critics are, his belief is well founded.

• • •

Meinig seems to believe that by writing down a lot of names of places and people, a story is effectively told and insight is conveyed. In *The Shaping of America*, it is mind-boggling how little is said *about* the places and *about* the people who are peopling America. Among hundreds of examples that might be cited is the following: "Missionaries had penetrated the barrancas to make contact with all the peoples of the Sierra Madre Occidental, and an occasional trader might make his tortuous way between the Pacific Slope and Parral or Chihuahua but there was no regular trafficway across that formidable barrier anywhere north of Durango."[10]

This information is potentially valuable if we know or are told something about Parral and Durango and the peoples of the Sierra Madre. But without a good bit more context and explanation than Meinig provides, it is all just one more mind-numbing "fact" (if that), at best loosely connected to his relentless hunt for yet more geometries, the big and the small political and social morphologies of unfolding America across five hundred years.

The very nature of Meinig's grammar compounds these problems. Rather than fill his sentences with detail that makes for understanding, Meinig repeatedly reverts to abstractions and dandified adjectives and adverbs that in fact convey very little information at all. For example, he begins one paragraph as follows: "As for expansions in the Old Southwest, there was a steady extension and thickening of settlement in Kentucky and Tennessee, states that grew by 342,000 in the first decade of the new century."[11] What does he mean by "steady extension and thickening of settlement"? That, for example, pioneers initially claimed five or six acres of land as they moved farther and farther into Kentucky and Tennessee, and then later settlers claimed plots of land roughly midway between these first pioneers? I don't know, and here as so often in these books, Meinig never stays put long enough in one place—despite the length of these books—to provide this kind of richness. It all goes begging, again and again in the name of defining lines, curves, shapes—geometries—if not on his maps, then all through his text.

Meinig, it would seem, who was more or less a contemporary of the imaginative Bill Bunge at the University of Washington as a graduate student, became infected with the same morphological or geometric bug that blunted Bunge's otherwise considerable vision.[12] It is not that Meinig doesn't do more than talk around and behind and about the unfolding political and social geometry of America, but rather that he is so preoccupied with geography as shape, form, position, and movement that he both tells a boring story and ignores or gives short shrift to so much else that, from a geographical perspective, is so much more important.

Sadly, Meinig seems to be very much another victim of the "spatial revolution" in geography. Like so many geographers, and a great many much younger than he, he never understood that "space" is usually a third- or fourth-order variable. It surely does matter how people are spaced, how large or small spaces are, how people move from one space to another, and so on. But these patterns and constraints require an in-depth analysis of matters that are fundamentally social, economic, and political in nature. To the extent that Meinig gets around to these processes, they are often buried if not lost in the wash and waves of his overindulgent spatial fix on history and his insipid "theory" and modeling, all to the detriment of so much else he could have written about.

This crippling spatial virus that predominates Meinig's vision has come at the expense of doing nothing at all with ecological matters, anything beyond the most perfunctory treatment. In a lengthy mea culpa, Meinig reaches for his rights and what they mean. His work is admittedly "idiosyncratic."

> I have focused on a few themes of strong personal interest and virtually ignored others of intrinsic importance. For example, I have not paid close attention to changes in the land with reference to environmental and ecological matters, and my intermittent mention of crops and commerce and of center-periphery relations barely hints at the kinds of geographic analyses of economic activities and systems relevant to historical interpretation. My emphasis on social and cultural patterns reflects not only what I enjoy learning about areas but, more important, my conviction that the geography of such things has been seriously undervalued in descriptions and assessments of the United States.[13]

Meinig certainly has the right to do whatever he pleases, which is quite different from a judgment about whether he possesses a large or a small vision. In my estimation, Meinig's "spatial" take on America to the exclusion of so much else that he utterly ignores is the very kind of perspective that has contributed in a major way to firmly establishing geography as a third- or fourth-order discipline.

The following entries are absent from the indexes in both volumes 1 and 2 of *The Shaping of America:* diet, family, demography, children, prostitution, drinking, crops, taverns, health, and food. And there are many more missing entries I might've included. (I don't really care whether Meinig personally prepared the index; as author, he's responsible for it.)

No one has to look at B westerns to learn that America's frontiers, like frontiers everywhere in the world, have had skewed sex ratios, that men have needed their taverns and their alcohol and sexual outlets for the women who initially were few in number. Alcohol and prostitutes didn't make America, and America would have been settled without them, but they sure made it easier, they sure seemed necessary at the time, and they often led to more than polite conflicts between those overabundant young frontier men and churches and reformers, wives and wives-in-the-making. You learn about none of them in Meinig's sprawling story.

What, anyone of mild curiosity might ask, was the average size of a Yankee family, a seventeenth- as opposed to a nineteenth-century family, an eighteenth-century rural as opposed to an urban family? How did health conditions change with the peopling of America? What—if Meinig were to give us just one or two examples—was it like living with a mouthful of rotten teeth, persistent skin rashes, or eyeglasses that were barely serviceable? What, beyond the vaguest kinds of generalities, could be said about family values, women's values, the education of children, and how all of it varied through time and from place to place as America slowly became the America as we know it today? On all of these matters, Meinig has nary a thing to say, and it's not surprising given his maniacal preoccupation with geography as space.

Geography as an academic discipline has a long history of being awfully insecure about its image, about what it's up to, about why it is so often ignored in universities or in this and that listing of disciplines that matter on this or that question. One of the ways in which geographers have made a mighty effort to deal with this insecurity is to look for reassurances from people in other disciplines that what they are doing is good, worthy, and indispensable. So Meinig, not surprisingly, wants the kudos of historians. In fact, he has always wanted their accolades more than he has wanted whatever he might get from his colleagues in academic geography. At least it's not a disease that he uniquely has.

But however much a few historians have liked some of Meinig's work, the insecurities remain. Meinig betrays this insecurity by returning repeatedly to that word *geography, geographic,* or *geographical.* His descriptions, then, can't just be straightforward descriptions in good, plain English of what happened or what is; they have to be identified as being "geographical." Some examples:

"Geographically, we can follow the shift of Mediterranean seafarers."

"Over many decades the geography of contact between Europeans and Indians. . ."

"Anyone who pondered the geographical dynamics of this encounter on this continental scale in 1750. . ."

"An important feature of this geographical pattern was the almost complete lack of lateral connections between these diverging northerly extensions."

"Such continuities in human geography—concordant or duplicative patterns—are most likely where there already exist in the imperial province obvious centers and trafficways."

"The character and degree of such geographic changes can vary greatly with specific circumstances."[14]

In virtually every case, the word *geography* or its variant can be eliminated, and not only is nothing lost in meaning, but the sentence is cleaner and clearer. It is clearer because *geography* is such a large and all-encompassing word that is it hard to know exactly what Meinig or anyone using the word as he does means. I doubt that in most cases Meinig himself knows what he means.

Meinig has insecurities (or should I say haughtiness that goes beyond all reason?) of a different sort, and they are found in everything he has written. He has a poor sense of understatement; he loves to find "roots," "fundaments," "essences." He wants to root and reinforce a building not only against the next two major earthquakes but for all time. He resists describing something with a straightforward statement; he wants to pound it into our heads, as if we are too thick to understand that historical events are contingent, or—yes—necessarily follow in some cases given certain axiomatic prior conditions. Either way, Meinig pummels his readers repeatedly with "certainties," "inevitabilities," or—what is somewhat different—*his* certain or superior knowledge of what had to have happened or could not have happened. A number of words and phrases that he uses repeatedly demonstrate this tendency. But nothing quite so consistently does so as his use of *of course*. For example:

"But *of course* discovery led to visions of conquest and exploitation."

"Yankee merchants were *of course* much involved in the slave trade."

"This is *of course* a promotional effusion typical of the company's literature."

"And *of course,* some of the Indian nations therein had been routine participants in American and European commercial systems for generations."

"Military occupation of the Mexican capital and other strategic places gave American leaders the privilege of considering whether to take much more territory and there were strong pressures to annex Tamaulipas, Nuevo Leon, Coahuila, Chihuahua, Sonora, and Baja California. *Of course,* there was no assurance that such areas could have been obtained." [15]

There are, I'd guess, a couple of hundred examples of *of course* in these books. And just about all of them are unnecessary and off-putting, indicative of someone either trying too hard to anticipate the exception raised or too eager to avoid being accused of saying something trite.

Meinig does the same thing with other words. A few quick examples suffice:

"Thus the *really* firm beginnings in America date from Ovando's. . ."

"Thus the America created out of this northwest sector of Europe was *necessarily* more invention than extension."

"But we *must fix* in our minds on the simple but long-neglected truth that Indians were *critically* involved in the creation of every European colony." [16]

We all make this mistake that I'm accusing Meinig of, and I'm no exception. The issue is how often. Often enough to slap the reader in the face and eagerly want to ask, Should we *believe* this guy?

A good bit of what I have criticized Meinig for, and especially the important charge that his books are, at best, mediocre narrative history, could have, and would have been, caught by a good and strong editor. That editor might have been anyone from Meinig's wife to one of the many academic geographers and others he cites in his acknowledgments. Or it might have been a good editor at Yale University Press, and Yale has some quite good editors, as I know personally from experience. [17] Regardless, Meinig didn't have that good editor for these books, and his *Shaping America* books are much the worse for it.

Why didn't Meinig get good editorial help along the way? It's hard to know without more information than I have, but one quite plausible reason is that he had far too much success too early in his career, and he managed over the years to surround himself with "friends" that didn't have the courage to say to him what they probably said all the time to undergraduates when grading their term papers. Unfortunately, "big" men (and Hemingway would be just one of many examples) are so big in their own minds and in the minds of friends that they are well above and beyond that part of the intellectual firmament where hard-nosed constructive criticism would have made them much better.

Finally, it is worth noting that after the first volume was published, Meinig remarked to an academic friend, apparently with some pride, that his editor at Yale had not changed so much as a single word. Nothing is so blind as the pride of fame.[18]

If we are willing to give the riches of a twentieth-century American city to someone who has taken on a task of the sort Meinig has, then we should be equally willing to exact riches of this magnitude from whomever informed the world that he wanted to climb the biggest mountain but never got much beyond Base Camp 1. In that category of mountain climbing to which Meinig has aspired in *The Shaping of America*, he has not gotten anywhere near the summit. He certainly has had the energy, but he has sorely lacked the open and expansive talent that is also required.

Part Three   ⑥   Misbegotten Landscapes

"What is essential is invisible to the eye," the little prince repeated, so that he would be sure to remember.

    —Antoine de Saint Expery, *The Little Prince*

Richard here and there

Coconuts on a Lava Flow
in the Chiricahua Mountains

> The toughest and most difficult occupation in the world, in my
> opinion, is to play the part of a king worthily.
>
> —Michel Montaigne, "Of the Disadvantage of Greatness," in
> *The Complete Essays of Montaigne*

CARL Sauer has been characterized as the most important cultural geographer of the twentieth century, and a scholar of the first rank.[1] Clarence Glacken, one of Sauer's colleagues for many years in the Department of Geography at the University of California at Berkeley and himself widely acknowledged as having written one of the finest scholarly works by any geographer, claimed that cultural geography in the United States *was* Carl Sauer.[2] He was not alone in his assessment; others have made even larger claims. Bob Callahan, for example, claimed that at Sauer's death in 1975, "the American people as a whole lost one of the most articulate scholars this century has yet to produce."[3] Even at midcareer, Sauer was referred to as that "Great God Sauer, sitting serenely beyond the Sierra Nevada."[4]

Among various traits that we ascribe to scholars is careful attention to the gathering of field data and to the written record, or, more broadly, the available facts, whatever their provenance. We do not expect scholars to purposely ignore what others have found on the subject of interest; on the contrary, we expect them to aggressively look for what others have had to say. We also expect that scholars will make every effort to weigh all available evidence before arriving at a conclusion, no matter how small or large. And, not least, we expect scholars to have a discerning critical faculty, to know how not only to weigh evidence but also to evaluate inferences that follow from the evidence.

David Stoddart, a geographer and geomorphologist, has gone to some length to examine Sauer's academic life as a geomorphologist, concentrating on the period of Sauer's life that lasted until the mid-1930s when he began seriously to turn his attentions elsewhere.[5] Sauer was then in his midforties, so the geomorphic period of his career was considerable indeed, and there's every reason to believe that it was formative in how he approached his research. It is worth highlighting several points that Stoddart makes, and then developing the arguments with further evidence from Sauer's long career that extended well beyond geomorphology.

When Kirk Bryan, about whom Sauer once had a high opinion as a geomorphologist, criticized Sauer's lack of attention to detail and complexity in his study of the Peninsular Range, Sauer not only was "deeply annoyed" at the criticism, but also said of Bryan: "[T]he more I see of [him,] the more I think he has the bumptiousness of the half literate." He further added: "Bryan seems to have undertaken to suppress all observing and thinking in this country which does not follow him or his prophets, and believe me he is a long way from having the skill or the knowledge necessary to the success of such censorship."[6]

Stoddart comments that "here Sauer adds Bryan to the Berkeley demonology of [William Morris] Davis." He might have added that the best way for Sauer to have refuted Bryan would have been to directly address what was wrong with Bryan's argument, point by point, or to go into the field again and come back with more data to show just how wrong Bryan was.[7]

In 1926, Sauer traveled to Baja California with some of his students. There they looked at changes in vegetation, and such landforms as sand dunes, granite outcrops, and terraces. Sauer was fascinated with sea-level changes and Indian shell heaps, and he thought there was a correlation between the two. As one of his students noted, "We argued some about the sea level changes and Professor Sauer was very reluctant to accept an alternative view."[8]

In Sauer's study of the Chiricahua Mountains in southeast Arizona in 1929, he came to conclusions concerning faulting for which there would prove to be no support, conclusions that Stoddart would decide were "*again* a priori and impressionistic."[9] A little later, Stoddart adds:

> The Chiricahua paper was not one of Sauer's greatest achievements. Indeed, not only was it an unplanned investigation intended to rescue a Mexican field season, but one has the distinct impression that most of his observations were from the railroad car. Certainly [Sauer] fails to mention

the most spectacular geomorphological features of the area, which now form the basis of the Chiricahua National Monument. It was a study characterized by all the faults of deductive reasoning attributed by Sauer and [John] Leighly to W. M. Davis.[10]

When Stoddart wrote this, he may well have been aware of another "railroad car" charge against Sauer, for one of Sauer's students, Robert West, leveled a similar "eye-to-facile-conclusion-from-the-railroad car" accusation against Sauer, here with regard to jumping to unwarranted conclusions shortly after Sauer arrived in Chile in January 1942. In one of his very early letters from Chile to the Rockefeller Foundation, Sauer wrote: "The one valley we've seen, cutting through from Andes to sea, the Aconcagua, is a torrent-swept waste of cobbles and sand." Robert West, in a footnote, had the following to say: "Sauer's impression of the Aconcagua Valley as a 'torrent-swept waste of cobbles and sand' was probably gleaned from his observations from the train window; for much of its route between Valparaiso and Santiago the railway follows the bank of the Aconcagua River, whose channel is indeed a 'waste of cobbles and sand.' The valley itself, however, is one of the heavily cultivated and productive areas of central Chile."[11]

In 1934, Sauer began working for the Soil Erosion Service as a consulting geographer, and in that year he commenced fieldwork in Polacca Wash, a thirty-mile-long channel, or arroyo, in northeast Arizona. He not only did not see the need for detailed mapping of geomorphic units in the channel, but also did not have any plans for bringing along measuring equipment; it would be the concern of others. On this issue, one of Sauer's persistent critics, Kirk Bryan, remarked: "No stream gaging or measurement of run-off is to be used. Now just what does this mean? Either I am very old-fashioned and even out of date [he was one year older than Sauer] or it means nothing and cannot accomplish anything." Bryan then went on to recommend that Sauer be removed from this research effort, "before," he said, "too many young scientists have been ruined by stupid programs laid out by men of inadequate scientific knowledge, training, and ethics." Stoddart adds that this statement to Isaiah Bowman was "intemperate and unworthy . . . [of] someone with Bryan's reputation."[12]

Intemperate, perhaps, but otherwise appropriate given Sauer's impressionistic, forget-the-hard-data approach. Was Sauer "shabby" if not "parochial," as Peter Gould has charged? Surely, one way of judging an academic as parochial is when there is an unwillingness both to be open to new ideas not your own and to change in the face of evidence brought to bear, especially by someone who has given more thought to the issue. Sauer needed to address criticisms, do more than merely see a place such as Polacca Wash as one of "evil omen" larded with "personal jealousies."[13]

Whether other geomorphic formations and the geographers working on them brought similar images to Sauer's mind is unknown, though it seems that Sauer had a problem dealing with hard evidence that ran counter to his beliefs. Sam Dicken, one of his students, has said:

> My biggest argument with Sauer occurred after our return to Berkeley. He had asked me to run a traverse down one of the arroyos near San Fernando. I made a sketch map showing basalt at one point, marked 'bs' and alluvial deposits on the narrow flood plain. When I saw the manuscript of the San Fernando paper, the basalt had become a lava flow and the alluvium a lake deposit. I protested that the basalt was in the form of a dike and was fine crystalline and that the alluvium was coarse sand. I protested in vain. It appears in the paper as a lava flow.

On this account, as on others, it looks like Sauer was one poor scientist, or scholar for that matter.[14]

Robert West, another Sauer student who has written about Sauer's fieldwork in Latin America, put it this way: "Many of Sauer's associates thought that he often arrived at sweeping conclusions too quickly from field observations, and on occasion was reluctant to change his views, even in the face of contradictory evidence." Further on, West writes, "Often [Sauer] was wrong and his students knew it. Sometimes Sauer would concede his errors; more often he did not."[15]

And then there is the telling case of Henry Bruman, one of Sauer's many admiring students. Bruman had a long-running dispute with Sauer about the origin of the coconut. He worked on this issue for his dissertation and published subsequently on it. Unlike Sauer, Bruman became convinced that *Cocos* as a genus and cultigen came from the other side of the Pacific, that it was not a staple among any aboriginal groups in the Americas and could, in fact, be found at only a few locations along the Pacific Coast, probably through a casual introduction from across the Pacific. Bruman published three articles on the coconut prior to Sauer writing about its American origins in the *Handbook of South American Indians*, which was published in 1950. As early as 1944, Bruman exchanged several letters with Sauer on the disagreement, but Sauer would not budge on his position.[16]

Rather, the letters show Bruman giving detailed reasons for his position on the coconut and Sauer, much of the time, appealing to authority, to Bruman's credentials, and to what was not known about the coconut.

When Bruman was working for the Institute of Social Anthropology at the Smithsonian, he wrote to Sauer: "I am frankly astonished that you still accept O. F. Cook's idea that the coconut is native to this hemisphere and that it was fashioned into a cultigen in the New World. Cook's case has been pretty well demolished by now." Bruman gave four reasons for his position, including the claim that "if the coconut was fashioned into a cultigen by American Indians, why is it that the early chroniclers say not a word about the use of coconuts by Indians?" He ended his argument with a plea to Sauer: "I do hope that you will reconsider your remarks concerning the coconut, since the *Handbook* statement will probably be widely accepted in the future." [17]

Sauer responded by claiming that Bruman was not being "scientific," that his position lacked "objectivity," and that he was being as dogmatic as Cook yet did not know as much as Cook. He added, "Cook has been for very many years a student of the taxonomy and ecology of palms; and I made inquiry among botanists, and they agree that Cook knew more about New World palms than anyone else." On the chance that Bruman might not appreciate the significance of this appeal to authority, Sauer added, "Incidentally, of course, you know that Cook is one of the few older botanists who has an understanding of the problems of genetics." [18]

But there was yet more, for Sauer ended this part of the exchange with an admonition, one of the sort that others might well have sent to Sauer in another day in regards to some of his geomorphic research and even to what he would later write in *Agricultural Origins and Dispersals*. He wrote: "I don't think that rushing into a terribly complicated question like this with some knowledge of the literature and very little of the habits of the plant, and arriving at a sharply defined position, is going to help you or science." [19]

Bruman responded by saying that he could see where a few corrections were in order, and that some of his sources were less than certain about what they had found. Still, on the whole, he stood behind his reasoning on the origins of the coconut, reasonably and openly concluding, "I feel pretty confident that my conclusions are justified on the basis of the available evidence. If future developments necessitate a change in viewpoint, I will be happy to recant in print." The latter is a concession rarely if ever heard from Sauer. He was simply not someone to admit to error in any venue.[20]

Still, this response from Bruman in hand, Sauer was not satisfied that he had put Bruman, the student—who usually referred to his mentor as "Chief"—in his proper place, and that he understood that now was the time to change his mind or back away from the research. Six days after he received the latest Bruman letter, Sauer wrote, rather ironically, "You're still over-confident and putting your bet too heavily on what seems to you to be authority, and you are trying to win a debate rather than investigating the status of knowledge and the uncertainty of evidence." In this exchange, Sauer did give reasons for his position, but in the end, perhaps uncertain that they were sufficient, and apparently eager to keep Bruman quiet on the matter, he downplayed Bruman's abilities and again returned to authority and, as it were, inadvertently, his own shortcomings.

> I don't think that you have gone thoroughly into the problem. I should feel happy if you were starting in to contribute to the geography of palms and to raising certain problems in their distribution and in their modification under man's care. That would contribute in time to science. You are not doing so by having already the courage of your convictions in a field where almost everything still needs to be discovered. It is not a question of your having confidence in your reasoning powers. W. M. Davis was a great arguer, but he ceased being an investigator early in his life.[21]

Among other things, Sauer could not appreciate the irony that if Bruman was not doing science because too little was yet known about—in this instance—the coconut, then surely Sauer was also rarely doing science.[22] (From the context of the exchange, there is little doubt that Sauer saw himself as a scientist, and Bruman as merely aspiring to this lofty category.) Nor could Sauer see that in less than eight years in *Agricultural Origins and Dispersals* he would be putting a lot more speculative and poorly understood ideas on the table with regard to domestication than Bruman had.

Whether Bruman was right or wrong was, I think, less important than the fact that he engaged the issue of coconut origins with something of an open mind, with more of an appeal to evidence and less to authority, and he was certainly much more willing to publicly admit error if proven wrong.[23] More than forty years later, the heat of the disagreement largely forgotten, Bruman saw that he had not been alone when taking issue with Sauer. Other students had been treated similarly.[24]

As for the issue of coconut origins, it now seems that Bruman was right, "for no one today argues for anything but an Indo-Pacific origin of this pantropical plant," even including Sauer's son who took issue with his father on not only the coconut but also the legume *Canavalia*. Working in his son's favor was a willingness to do the tedious scientific work that increases the likelihood that results will be less subject to disputation in the future.[25]

Sauer was single-minded on other issues, and on one in particular— namely, the overriding importance of diffusion as opposed to independent invention—it would prove much more debilitating. He failed to see that most crops have been domesticated more than once; innovation was much less rare than he claimed. Diffusion was just not as significant in explaining world agriculture as he asserted.[26] Likewise, Sauer may have gotten stuck in a rut when he made an assumption about productive efficiency, a cul-de-sac that prevented a more complex view of agricultural practices. As Gade has noted:

> Sauer constructed a chronological sequence partly by making idealist assumptions about productive efficiency. However, agriculturalists also may have sought to increase diversity for reasons that have nothing to do with utility. Eventually usefulness became recognized, but even then subsistence security, not productivity, could have been the prime motive for coaxing wild plants into crops. That could explain why inefficient cultigens continue to be grown in spite of their low productivity and the time required to process them.

And there was a good bit more that Sauer was wrong about when it came to agriculture.[27]

In addition to being parochial, as I have used the word, Sauer could really hold a grudge when someone took and held to a contrary position. Thus, Bruman noted that his disagreement with Sauer over the origin of coconuts created a gulf between them that lasted more than a decade, a period during which Sauer, on one occasion, obviously snubbed Bruman after he had given a talk at Stanford University. Not only had Bruman crossed Sauer, but he'd also had the audacity to work on a topic, Bruman concluded, that was not to Sauer's liking! Yet Bruman, ever the loyal student to the man and the legend, tried to put the best face on some unenviable traits: "Sometimes he would let his enthusiasms run away with him, and sometimes he hung on to his notions too long when the contrary evidence was stronger, but these quirks just showed he was human; it was good for the rest of us to see that the master also had a few limitations."[28]

Sauer's stubbornness in the face of new evidence went well beyond his dispute with a former student. In 1943, he submitted a manuscript to Julian Stewart for inclusion in the *Handbook of South American Indians.* Titled "Cultivated Plants of South and Central America," the paper assessed what was known about virtually the whole range of crops grown between southern Chile and the Rio Grande. Before submission, the manuscript was critiqued by the botanist Edgar Anderson and other scientists. Some minor and some major errors were corrected.[29]

Originally scheduled for publication in 1945, the *Handbook* was delayed until 1950. During this seven-year interval, several studies were published that should have changed Sauer's paper in important ways: on the classification of maize and its age in the American Southwest and the implications for its age in southern Mexico, on the potato's pre-Colombian presence in the New World and its taxonomy among the Aymara, on evidence of early agriculture in western South America, and on minor Andean crops. But as Daniel Gade notes in *Nature and Culture in the Andes,* "none of the above findings got into Sauer's plant chapter."[30]

Did Sauer ignore some or all of this work because of his refusal to budge from what he believed and had put to paper? Or had Sauer not kept up with research that was relevant to his interests and continuing interests, as would be evident in 1952 with the publication of *Agricultural Origins and Dispersals*? The latter issue is more about bad scholarship than intransigence, an issue every bit as important to an assessment of Sauer, the often-noted "scholar."[31]

The words *scholar, scholarship,* and *fieldwork* are deeply ingrained in the lexicon of the academy. To be called a scholar or to have a book that one has written called a work of scholarship or to have done fieldwork sets one apart from journalists, travel writers, essayists, and most intellectuals who write nonfiction. Not only are all of these latter groups doing something different, but that something different is almost always considered inferior to the work of a scholar.

But if there are notable differences in theory between scholars and others, the differences in practice are very often a great deal less than all but a handful of academics are willing to admit. For every first-rate scholar in the academy there are a half-dozen or more laying claim to this rubric who are producing work that is not only not more reliable, more insightful, or more objective than that done by journalists, essayists, and others outside the academy, but in many cases simply inferior as well.

What, in fact, is often passed off as scholarship is, on the whole, a messy and unpalatable soup of dubious truths, salty half-truths, and raw opinions, all of it given the weighty air of authority via endnotes, footnotes, and long bibliographies that no one cares about. So-called fieldwork is every bit as suspect. The vast majority of academics who claim to do field-work do little more than travel to a foreign place; spend tiresome days in li-braries, archives, and government offices in search of someone else's prejudices and half-truths; and then, when they are out wandering through towns, cities, and country landscapes, embrace the ill-conceived conceit that their eyes reveal arcane cultural codes and long-lost secrets that other mortals not similarly "trained" cannot possibly see. When it comes to talk-ing to people, and getting aggressively tough as you very often must to get anywhere near the truth about much of anything, the great majority of aca-demics (there are some quite notable exceptions among anthropologists) are rarely more than rank amateurs. They do not have the basic snoop skills of a decent journalist, and the ones that do are so afraid of the smallest dan-ger or the toes they might step on that they might well have been better off staying at home and just guessing what they might have found out by doing fieldwork.

Little of what I have just said can or will be openly admitted by more than a very small handful of academics I have ever known, and for the very good reason that to make such admissions is not only to take away the "specialness" of what academics believe they are doing and to be forced to admit that they are trafficking in prejudice every bit as much as journalists and the others they so openly disparage, but also to expose openly the fragility and hollowness of their so-called scholarly research, the mundane-ness of their much touted and highly overrated "fieldwork" skills, and the numbing ordinariness of so much of what they write and about which they make such outrageous claims of insight, originality, and the like. The acad-emy is not just thoroughly elitist, but also full of more empty, baneful, and corrupting conceits than anyone has yet managed to list.

Put differently, exposing most academics for the flabby intellectual frauds and lousy field-workers they are is akin to publicly exposing a macho male's manhood, and in a way that could not be more embarrassing. No sin-gle accusation in the academic world is more damning than this one: You are a sloppy scholar. Worse yet, to be told that what one is writing is not scholarship at all is the ultimate insult, a curse not unlike that of having your medical license permanently revoked for unforgivable incompetence.

With a few minor exceptions, there simply is no place within the academy for journalists, for essayists, for the ones who would dare proclaim that they are not doing scholarship. Indeed, rare is the individual in the academy who even entertains thoughts of advertising himself in these terms. And, of course, it is with considerable reticence that any academic accuses another academic of not doing scholarship. He may say that the purported scholarship is not quite up to his standards, or that the methods leave something to be desired, or that he disagrees with the premises, the conclusions, or the form of the analysis, but he will not openly charge that it is not scholarship—not, at any rate, without considering the considerable personal and social consequences.

No one, to my knowledge, has ever said that Carl Sauer was not a scholar. Rather, the widespread, almost universal consensus has been that he was not just a scholar but also among the very best. And not just in geography. To be sure, there was the occasional letter that Sauer received that questioned why he had not cited so-and-so, or one that wondered where he got his information. But this kind of questioning, whether of Sauer or anyone else, works from an assumption of innocence. Something was innocently missed, in haste or through inadvertent neglect, not through any kind of purposeful deviousness or consistent sloppiness. There was even the occasional Sauer student who was taken aback, even angered, that his mentor had failed to give credit where credit was due, or that he purposely ignored contrary evidence. These failings, I would bet, were seen as minor transgressions, easily forgotten slips, nothing to fundamentally scar the esteemed scholar and his image.

My accusation is that Carl Sauer was, at best, a mediocre scholar. At his worst, and much more frequently than admirers would be willing to admit, and the published record shows, Sauer was not a scholar at all. He was, more than anything, a provocative essayist in academic cloth who documented just enough to keep the hard-nosed critic in his den. Sauer the "scholar" was lazy; good scholarship, like good science, is a lot of drudgery, and this kind of effort for someone of Sauer's focused self-importance was never on the agenda.[32]

Even regarding his famous small monograph *Agricultural Origins and Dispersals*, some of his strongest admirers, and Sauer himself now and again, referred to it as an "essay." Here, however, the term, I venture, was used as a form of flattery, meant as much as anything to highlight its broad and far-reaching ideas, some of which, as time would show, would prove dead wrong. Whatever, *Agricultural Origins* then, and now, has been seen as scholarly, fundamentally. That it was not, that it was indeed no more than a quite speculative essay, has not been sufficiently noted. Perhaps had someone done so, there might have been a serious question raised about just what Sauer was really doing.

There is nothing at all wrong with being known primarily for generating ideas, for hypothesizing all kinds of new and different relationships and connections that are dimly perceived, if perceived at all, by others. A person with such qualities is of immense value in and outside the academy, for the simple reason that so many people in and outside the academy are just plain dull, and bereft of ideas of any sort. They are that 95-plus percent that Thomas Kuhn has referred to as normal scientists. They follow, they parrot, they test what others suggest is worth testing. In one sense, then, Sauer's admirers are right when they want to bestow on him an elevated place in the history of American geography. But Sauer's admirers and followers are wrong, very wrong, in equating this trait with scholarship, or in thinking that it does not matter how often you are wrong. It matters a great deal how often a scholar is wrong, for nothing is more difficult to get out of the minds of students and the legions of unthinking professors than that which has appeared in cold, hard print under the name of someone not only famous but also alleged to be a first-rate scholar.

Was Carl Sauer a good, nay, a great, field-worker? Certainly, he has always been characterized as a master of getting about in the field. But I believe that this claim may be every bit as misplaced and wrongheaded as the one concerning the quality of his scholarship. His field notebooks are just about the shabbiest I have ever seen (I have examined the ones that have survived and are in the Bancroft Library, the Sauer Papers); they would not receive a passing grade in an undergraduate course in which the assignment was to walk around for a week in a village and write down observations on what was discovered through casual conversations.

The widespread claim that Sauer's "fieldwork" found its way, in an integral and important way, into his research is just not entirely credible. Virtually all of his work was historical, dealing with the distant past; although he did pick up quite useful archaeological and other information now and again in Mexico on his trips, he could have gotten a whole lot more out of fieldwork had he been much more systematic, had he not placed so much emphasis on what he thought he saw, had he not made more—much more in some cases—of his wandering and cursory fieldwork than was warranted. His "fieldwork" may have given him an edge over others in his understanding of aboriginal population numbers, a major concern; it may equally have given him confidences that were not at all warranted, as with his grandiose extrapolations from the present to the past and from local observations to ones relevant to large areas. Sauer might have been better off, as a scholar, and not so often wrong, had he foregone several of his wanderings in Mexico and instead spent the time in more careful analysis and documentation of the historical record.

Carl Sauer's disdainful attitude toward careful scholarship and careful fieldwork can be explained in part by his high self-regard and strong sense of Teutonic superiority.[33] These traits may well have given him the confidence to go where others would not go with ideas, to brazenly exhibit untested hypotheses as if they had firm truth status, and to provide a model of godlike behavior for those many students who could not stand on their own two legs and who needed a strong and domineering father figure to tell them what to do. The academic world, as anyone knows who has worked in it, is full of weak and fragile people who cannot distinguish the word *courage* from the word *cougar*. Alas, very often it seems to require the domineering, all-knowing presence of a Carl Sauer.

Lost in Lies about California's Migrant Labor Landscapes

> The truth? A half-truth? There was no way, I realized with a sudden pang of anxiety, to gauge this . . . sincerity.
>
> —J. Morrow, *City of Truth*

I didn't go to Berkeley, and therefore I never had an opportunity to have Jim Parsons as an adviser, or to see him on a more frequent basis than I did. Through a series of fortuitous circumstances, however, beginning in 1968 in Costa Rica, I got to know Jim rather well. Whenever I found myself in the Bay Area, I'd make an effort to call him or try to find him in his office on the Berkeley campus. Over the years, I wrote to him often, and I heard from him often—he was one of those rare dependable correspondents. Over time, he came to see me (in his words) as his "adopted Berkeley son," and beyond this description, I think it fair to say that we both had quite a bit of affection for one another, not unlike what he developed with some of his favored students. There was—to be sure—a good deal we didn't agree on.

One of our major differences was that Jim didn't want anything to do with controversial topics or academic controversy—at least, to my knowledge, until his lifelong home at Berkeley was under final assault by colleagues whom he had, I think, just plain come not only to distrust but also to loathe.

My first encounter with Jim's attitude toward controversy came fairly early in our relationship, in Costa Rica, when I wanted to get involved in a research project on the implications of that small country's obscene population-growth rate. On mentioning this interest to Jim, he screwed up his face and said, Why would you want to do something like that? His tone was clearly censorious. A variant of the attitude would surface many years later

115

when I wrote to him and asked for more than moral support when the AAG censured my book *The Immoral Landscape.* He wrote back to me and said, "I just don't have a taste for those kinds of battles. Sorry." It took me quite a while to forgive him—such was my hurt that he wouldn't lift a finger in my First Amendment fight (but then, in fairness to him, neither would seven other past or future presidents of the AAG). But forgive Jim I did, and before long we were back on course in our oddball kind of father-adopted-Berkeley-son relationship.

These few personal revelations about Jim Parsons might seem to be going nowhere were it not for the fact that I just finished reading Don Mitchell's *Lie of the Land.* It turns out that one of the larger claims that Mitchell makes for his book (in the conclusion) is that Parsons felt that "the type of history [Mitchell has] told is unnecessary for understanding and appreciating the beauty of the Central Valley landscape."[1] To Parsons—as Mitchell would have it—the California landscape was all about aesthetic enjoyment and invisible and irrelevant workers who were simply "colorful additions" to the Central Valley. Mitchell, by contrast, is serious; none of this playing around, Parsons kind of nonsense. Mitchell is exercised by what he calls the "ugly" side of the history of California agriculture—the unconscionable exploitation of labor. Telling this story in stuttering fashion for the period 1913 to 1942 is Mitchell's contribution to knowledge. And not much else from what I can tell.

Mitchell is smart enough to not claim any theoretical advance, because there isn't any. And he's even smart enough not to claim that he's the first to look at this issue, which he isn't. He might've claimed that by virtue of mining nearly one hundred boxes of old California Commission of Immigration and Housing records he's added some new details to the story, and maybe he has, but if there's anything big or noteworthy in these new details, then he's not telling us in his wrap-up. Nope—strangely enough—it's Parsons and his sterile "aesthetic view" and some pretty-picture book authors who never heard of Marx and the means of production that he's after by way of closing his skip-hop-and-miss story.

Mitchell sets Parsons up for his final-round trump-card killing by going on at length about him in his opening chapter, "California: The Beautiful and the Damned." For what I'll have to say about Mitchell, it's important to quote him in full.

> In the words of geographer James Parsons, the [California] landscape "is morally neutral." As neutral, both people and landscape may be trans-

formed in their mutual encounters, but the moral content of the landscape remains fixed and imperturbable. It just is. The landscape is thus often understood in two interrelated ways: it is a relict rather than an ongoing construction; and it is organic, natural, and aesthetic. In the first case, the landscape is understood to be immutable at least in terms of the normal human life span. Rather than being molded directly by people, the landscape's immutability allows it to shape humans. In the second case, the landscape is something to be passed through and admired along the way.[2]

Mitchell has much more to say in this chapter about Parsons before he finishes him off: that he's "paternalistic," that his descriptions are "mere representations of the 'honest' diversity of the place," that "Mexicans and Filipino workers become curiosities to be gawked at," that the "landscape is purely a place of aesthetic wonder," and "that how it got that way is of little concern."[3]

Mitchell didn't go after the best writer on the subject he could find for a fight, just the best geographer with an abiding interest in California. It has long been well known that among geographers, Jim Parsons was one of the more astute and observant students about California. He traveled the state widely, he knew it well, and he taught at least one class on California at Berkeley for many years. And Parsons, very unlike a lot of other students of other ordinary landscapes, was a talker and attentive listener. In Costa Rica, where I first saw him in action, he was in furrowed fields talking to peasants before the rest of us had hitched our pants and found our brains. He may not have talked perfect classroom Spanish, but he found out all kinds of things he wanted to know and other things that he didn't know he wanted to know until he had digested what he'd heard. I have few doubts that when Jim traveled alone or with students in the state he so loved, he did the same with Mexican field-workers in the San Joaquin Valley, with wildlife biologists on the shores of the upper Sacramento, and with a whole lot of other people I wouldn't know how to classify. It's also no secret that Parsons was, after Sauer, probably the most important person in the history of Berkeley geography—if only to judge by the many Ph.D. students he produced: more than thirty. Then in 1986, well along in years but still hitting the road with fresh eyes, thinning hair, and an open mind, Jim wrote an essay for the *Geographical Review* titled, "A Geographer Looks at the San Joaquin Valley."[4] It was this out-of-war-zone piece that gave Mitchell a fit and caused him to go at Parsons with all kinds of off-the-wall accusations.

Although Mitchell doesn't extend his charge against Parsons beyond what he wrote about the San Joaquin Valley (to my knowledge), there's no reason to believe that Mitchell wouldn't also hold that Parsons was utterly indifferent to the decimation of the Sacramento's salmon population; that he couldn't, for the life of himself, "see" and understand how the Delta's blue-collar farmers were mindlessly destroying the Delta; and, to return to Mitchell's interest, that it just never occurred to Parsons that poor, illegal Mexican hunchbacks had, since the beginning of time in California's factory fields, been screwed over in more ways than most of us can imagine. Mitchell, after all, has characterized Parsons as "morally neutral," no better (nay, worse because he was a geographer with a trainload of students at that flagship school known as Berkeley) than the current round of pretty-picture book peddlers who can't see much below the enchanting pink-and-purple evening sun setting over a Modesto schoolhouse, no sweaty field hands in sight. According to Mitchell, Parsons certainly could never have done what Mitchell has done in *The Lie of the Land:* give us "a way of seeing that helps make sure those asparagus planters do not indeed simply dissolve into the foggy landscape they made" (the last line of his conclusion).[5]

I opened this essay by noting that Parsons had little taste for intellectual or other controversy, at least in my experience. What, then, can I say about Mitchell's charge in light of what I know? I have to confess that it's possible that Parsons was largely indifferent to the issue of labor exploitation by California farmers. By indifferent, I mean that although he may have been fully aware of how they were treated (and I'm sure he was), it wasn't his intellectual terrain because: (1) he felt there was nothing he could do about it (no less true of Mitchell), so why not do something else? (2) to the extent that there was a good story there (which is true, and he would have said so), he wasn't the best person to tell it (doubtful in my mind—if he'd wanted to); and (3) the talents he did have were best utilized elsewhere (oddly enough, probably right, notwithstanding what I just said).

But even if my characterization of Parsons is as good as any we might get from the people who knew him better than I did, what Mitchell charges is, I'm fairly certain, quite different from what Parsons was mindfully aware of and may well have talked about with students in his own quiet way but refused to write about. Even if he didn't say much to the people he traveled with when on the road in search of more tidbits—large and small about California, it's impossible for me to believe he didn't see the issues pretty clearly. I'd even bet he had enough stories of the sort that interest Mitchell to have filled a book or two. And to boot, the stories would have had a strong human element to them; they would have seemed real, with names and places given; they would have been accompanied by a Mexican or Filipino story and the social, personal, and economic tradeoffs that the exploited worker was making; they would have been set in the context of a real farm and the larger global economy; and . . . I can only imagine how many places Jim would have taken the story that Mitchell isn't aware of.

This defense is not an excuse for Parsons, but rather a way of explaining why I think he walked away from an "ugly California agriculture" story. In fact, my own commitment, much of the time, is closer to Mitchell's: social injustice. Unlike Mitchell, however, I do not have any interest in Marx, nor do I feel there is any need to bring him into the picture to understand how and why people are unjust to one another.

To reiterate a bit, I'm just not certain that Jim Parsons was—in mind—"morally neutral" toward the issues of rape and human pillage in the history of California agriculture; it was, I think, the kind of intellectual "fight" he had no taste for, any more than he had a taste for going after Costa Ricans for having too many kids, West Coast commercial fishermen for driving Sacramento coho and king salmon to near extinction, or helping me fight all those moral and intellectual nitwits in the AAG who killed a book I'd spent five years on. But then, in a quite real sense, the jury on Mitchell is still out; we need to see how tall he stands when he does some real fieldwork on the contemporary scene, and exactly what kind of bravery he shows when he's likely to lose some skin for fighting all those people in academia (his day-to-day turf) who daily screw students and faculty. If Mitchell doesn't stand up and fight there, then for my money he'll be just another breast-thumping, bourgeois radical living a comfy, middle-class life who takes his kids out for designer ice cream on Friday afternoons after his week-ending lecture on the profound injustices in capitalist societies à la Marx and his pigeon-toed followers.

What about the other charge against Parsons by Mitchell, namely, that he saw the California landscape, nay, any landscape, as unchanging, immutable, natural, organic, a relict, and the like? All of it is just plain hogwash. And for a straightforward piece of evidence, one need look no further than Parsons's exemplary doctoral dissertation on Antioquia, Colombia, to anchor my point. Like Mitchell's *Lie of the Land,* Parsons's late-forties Ph.D. was turned into a book, was eventually translated into Spanish, and is anything but out-of-date.[6] It is a book far superior to Mitchell's in every respect, and one that established beyond doubt that if Jim Parsons was anything, then he was most certainly a historical geographer, and one who saw landscapes as ever changing and always modified by human action. Without belaboring the point, on these many charges Mitchell is just plain dead wrong; he didn't do his homework.[7] He reduced the lifework of Jim Parsons to one article, akin to judging the literary career of Gabriel García Márquez by the fawning silliness he has written about Fidel Castro and the Cuban Revolution.

One of the tragedies of Mitchell's book is that he had a great story to tell (no matter how many times told), and he told it poorly. Good narrative history, like good literature, is like a good, uninterrupted dream. It's not littered with words in italics to let the reader know that he just *can't* miss this point and is *too* stupid to *get* it if not *italicized*. It's not a story continually interrupted by irrelevant quotes and asides to people who have little or nothing to do with the story. It's not a story well served by an introduction that pays far too much attention to twittering nonsense concerning this or that geographer (invariably one who, firsthand, in fact knows little about ordinary landscapes) who asserts that knowing landscapes is all about morphology or nature or culture or representation, ad nauseam. It's not a welltold story where the author genuflects to Marx and his offspring and then really doesn't know what to do with the sacred texts other than to use a lot of mind-numbing words that detract from rather than add to the palpable pain that with good, ordinary, everyday words he might've conveyed: about broken backs and too many herbicides floating around in the lungs and too many lice in the hair from living with too many other downtrodden "wetbacks" in the same one-room hovel.

I'd like to see Mitchell describe one of these "ordinary ugly California landscapes" up close. It won't be easy, which I can attest to personally. I've been physically threatened several times in Latin America when doing fieldwork, but I've never been threatened as often as I have when wandering uninvited through lettuce and melon and strawberry fields in border Texas or the Imperial Valley or Kings County, California. I don't know how often Jim Parsons went looking for Mexican informants in California's unfriendly carrot or beet fields. I do know that Mitchell ought to try it, and then write about it. There are plenty of good exploited-labor stories, and they're every bit as important as the one he tackled; he has a chance to make it all a lot more meaningful than *The Lie of the Land* if he's got decent eyes, sound ears, and a few language skills.

The preceding, by way of another digression, raises the what-if question: What if Mitchell had done his graduate work with a Jim Parsons rather than with a Neil Smith?[8] Well, if he had never encountered injustice and didn't know what it looked like, then opting for a Neil Smith was the right choice. That Neil Smith also fed him a feedlot full of nonsense about Marx, class struggle, and modes of production was only a huge minus in my book, and if Mitchell has any sense at his still young age, then he'll run like hell from this detritus, as fast as his bare feet will carry him through the Syracuse snow. If, on the other hand, Mitchell had a decent sense of what injustice was all about when he left San Diego for graduate school, and he had his mind set on pursuing a dissertation where he could examine injustice (of any old sort—the world is chockablock with it; there's nothing special about screwed-over field labor in my native state), then he would have been much better served finding his Jim Parsons. This Parsons he found might've raised his eyebrows once or twice at Mitchell, and he might even have mumbled something or other about looking at the origin of alien grasses, but Jim would have kept Mitchell's head clear of going-nowhere ideologue chatter. He would have gently shown him how to write the great seamless story he had before him. And he would have gently prodded him to write a much bigger, more complex, and more satisfying story than he did.

Knowing all that I know in retrospect, and knowing full well that I would probably prefer Neil Smith to Jim Parsons were I at real war and needed a gutsy body at my side to swing something substantial, I still would have gone to a Jim Parsons if I had wanted to learn how to write good and convincing history and geography—about anything. Good storytelling and the tools required to do it aren't everything, but had Mitchell known what engrossing and convincing narrative was all about and had he some honest writing skills, then he might've written a memorable book rather than one easily bested and just as easily forgotten.

After I wrote the above and was about to zip it into that underground tunnel known as e-mail to friends and a few wayward others, I got a feeling that, hard as I was trying, I wasn't giving Jim his fair due and I'd better go to the library and look up this San Joaquin Valley piece he'd written and I'd entrusted in interpretation to Mitchell. That little shit-detector voice that marches around inside my head told me that something wasn't quite right about the way Mitchell depicted Parsons. And so here's some of what I found.

Parsons's clear intent was to provide a large overview of the valley, touching on perceptions, settlement history, past and present environmental problems, internal place-to-place differences, the role of petroleum, the pressures of urbanization, and a whole lot more. In typical Parsonian fashion, the piece is beautifully written, and I would be proud indeed to have my name on the essay.

Having said the above, was Parsons, in this one article, just what Mitchell claims, or was Mitchell being dishonest? Yes, I do mean dishonest. Well, it turns out that under a major section titled "Interpreting the Valley," Jim identifies three principal ways of "looking at the Valley," and the second way, to which he devotes three long paragraphs, is "as a symbol of capitalism gone rampant, of all that is bad about profit-based, large-scale, labor-intensive irrigation agriculture." He notes that the Southern Pacific Railroad was depicted as the "Octopus," when "the railroad ruthlessly drove settlers off the land." And he notes that both Carey McWilliams (one of Mitchell's heroes) and John Steinbeck (whom Mitchell quotes at length in his book) continued this theme of ruthlessness in *Factories in the Field* and *The Grapes of Wrath,* respectively. Parsons continues: "Current critics are as likely as not to depict grasping corporations, made fat by subsidized water, exploiting hapless immigrant field workers in a system excessively dependent on agricultural chemicals, costly machinery, and the economies of scale—or getting bigger."[9] And there's more in this paragraph that sends the same message.

In the next paragraph, Parsons begins: "Enormous environmental problems confront the San Joaquin Valley; too much or too little water, water of the wrong kind and the wrong place, too much use of pesticides, herbicides, and excessively heavy machinery, land becoming compacted, too salty for cropping, fertile soils going irretrievably under asphalt." And Jim goes on, ending the paragraph with this sentence: *"These and other matters undeniably demand attention."*[10]

Mitchell, for his part, can't do much better than reduce all this Parsons-in-another-key to footnote 9 in chapter 1, which reads: "Parsons actually invokes the image of 'capitalism gone rampant' only to criticize those he feels dwell too much on the negative side of California agrarian development."[11]

Really? I forgot to mention that earlier in the essay, Parsons had no problem speaking of the valley as "cursed" "by poverty and arrogant wealth," nor did he shy away from paraphrasing J. B. Jackson's observation about rich men who have so effectively exploited the countryside.[12]

Mitchell gets into racism in his book (who doesn't when you're talking about who's screwing whom?), but when he comes at Parsons he characterizes him, as I've noted, as "paternalistic," and says that in his writing, "Mexicans and Filipino workers become curiosities to be gawked at," "the landscape is *purely* a place of aesthetic wonder," and the like. To these examples, I must apply one of my favorite words: *bullshit*. In the first paragraph of a section titled "Ethnic Diversity," Parsons notes that "when anti-Chinese sentiment intensified, they drifted to the cities."[13] In the same paragraph, he draws attention to the Nisei repatriates who were in World War II internment camps, which hardly sounds like a man strapped with racial or ethnic blinders.

This kind of detail by Parsons (and other points noted above) in an essay that *by design* is a synthetic overview is hardly the kind of stuff to be written by someone who is "morally neutral"—Mitchell's key piece of two-word evidence for hanging and quartering amoral Jim Parsons. Parsons's "morally neutral" statement needs to be put in context. He wrote:

> There is much to see and to wonder at in this magnificently complex, man-made countryside of the San Joaquin Valley. Yet there may be individuals who consider the idea of studying, much less admiring, the landscape of modern agribusiness to be repellent. There is no certain relationship between moral virtue and aesthetic values. The landscape is morally neutral. Is the valley less interesting, or its color and geometry less worthy of attention, because some of its harvests enrich soulless corporations, its landscape creations of the producers of nonunion table grapes or boycotted wines?[14]

Mitchell's answer to this question is YES.

Parsons's answer to this question is NO.

My answer to this question is NO. But my first gut reaction to Parsons's five-word sentence, "The landscape is morally neutral," was: Jesus, Jim, why did you write that damn sentence, so naked and bald and easily assaulted? My second reaction was: Jim, you're right, and the reason doesn't require a whole lot of explanation. The reason is that landscapes as visually known are indeed morally neutral. What are *not* morally neutral are the people who created, changed, altered, manipulated, screwed over, and exploited the landscapes.

It is true that to some people, landscapes include all these people who are not morally neutral too, so looked at in this sense—to include these people—no landscape that contains humans is morally neutral. But then this statement is true for every single landscape that has ever felt the imprint of humans. There is, thus, absolutely nothing special about the Central Valley and predatory agribusiness sorts in this regard. And—to push one edge of this argument a tad further—anyone not writing explicitly and at length about exploitation in some important sense of the word might be accused of being "morally neutral."

Jim Parsons was in love with landscapes in the material sense of the word. It is what he wrote about in his essay on the San Joaquin Valley and elsewhere, and it is what consistently captured his imagination and attention. It is the well-known "morphological" sense of landscape that caught Sauer's and Parsons's attention (he was Sauer's student) and generations of geographers right down to the present, including the likes of J. B. Jackson, Peirce Lewis, Wilbur Zelinsky, John Fraser Hart, John Jakle, and Terry Jordan. And even Don Mitchell.

For Mitchell, it is labor that is "producing" and "shaping" landscapes. For Mitchell, "re-representations" of landscape are the products of labor. For Mitchell, landscape is about "morphology" and a "view" of landscape. For Mitchell, "landscape is thus a unity of materiality and representation, constructed out of the contest between various social groups possessing varying amounts of social, economic, and political power." For Mitchell, "meanings are both posited in and developed out of the landscape's morphology."[15]

There is more to flesh it all out, but that is not my purpose here, nor am I interested in tackling this uncertain relationship between moral virtue and aesthetic values to which Parsons alludes. What is at issue is Mitchell and his book, and his charges against Parsons and the kind of geography he represented.

The accusations are patently unfair.

And Don Mitchell, in a couple of words, is—to judge by how he has treated Jim Parsons—a shoddy and dishonest scholar.

Intellectual Turf and Elitist Notions
about Invisible Landscapes

> One's self-satisfaction is an untaxed kind of property, which it is
> very unpleasant to find depreciated.
>
> —George Eliot

IN thirty years, I've written one book review, and that one, not an ordinary
book review, was published in 2000.[1] I haven't written book reviews because, on the whole, I've rarely felt I could be fair enough to the author, or I
haven't been willing to take the enormous amount of time to do them the
way they ought to be done, given how much time an author puts into a
book. Rarely, except in essay reviews, do editors give a reviewer enough
space to be fair and do justice to an author. But if I don't write book reviews,
I do read them, principally to see whether a book sounds interesting enough
to have the library send a copy to my university mailbox. Now and again, I
get fascinated more by the reviewer's tactics and self-promoting aims than
by the book under review, for reviews are often written for reasons other
than simply giving an honest account and critique of what to expect.

One of the ways in which reviewers use reviews is to settle scores for
personal slights, real or imaginary bad behavior, and ideologically incorrect
behavior. The reason for the score settling is never explicitly specified; editors not only won't allow it, but will also usually deny that score settling or
one-upmanship is even taking place.[2]

The score-settling exercise, as everyone knows, is easily done by sprinkling a little praise here and there and then leaving the unmistakable message that the book leaves a lot to be desired: the author lost his way, the
book is just plain awful, or the reviewer could have done a better job while
watching TV and tending the kids in the backyard.

Another way in which reviewers use reviews is to make it clear that however good the book, the reviewer is still king of the hill, if not with regard to the specific substantive matter at hand, then most certainly in the larger theoretical arena—where, of course, real points are scored for the hereafter.

The strategies for accomplishing these goals vary greatly, from pointing out major conceptual flaws and misinterpretations or misrepresentations of key sources to identifying major themes omitted and one or more glaring, if not monstrous, factual errors. Points scored have a particular poignancy if it's the reviewer who's the authority rather than some unknown pretender. To add a little malicious spice, the reviewer may even make the not very subtle claim that his remarks about what's in need of repair are little more than throwaways, not yet worthy of packaging to send off for publication.

There's more that could be said about the score-settling, one-upmanship side of reviewing and how much it really is about personal and social politics, but all of this poorly explored terrain is not my major concern here. At issue is the specific instance of Dick Walker's review of Don Mitchell's *Lie of the Land: Migrant Workers and the California Landscape* that appeared in the *Geographical Review*.[3]

Walker's review caught my attention for several reasons: (1) he's eager to let us know that when it comes to understanding these enormously important California agricultural landscapes, he, and not the younger comrade upstart Don Mitchell (they're both out-front Marxists), is master of the mountain; (2) just like Mitchell, he's taken an unfair (if smaller) whack not only at Parsons but also at so-called Sauerians; (3) he's making a great fuss about an awfully simpleminded idea; and (4) like Mitchell, Walker is not only wrong but also elitist in his insistence that California's agricultural landscapes are invisible.

A couple of socially and personally pertinent particulars need to be mentioned to provide context on Dick Walker on Don Mitchell. One is that not a scintilla of love was lost between Jim Parsons and Dick Walker while the two tried to live together in the same geography department at Berkeley. Parsons, who voted against Walker getting tenure, would have been happy to send him into permanent exile atop a ridged field in the San Jorge River of Colombia. Walker, no doubt, contemptuous of Parsons for his brand of geography as much as for his politics, would have been equally delighted to help Parsons get lost in head-high Kikuyu grass somewhere in outback Brazil. So it wouldn't be at all surprising if Walker tried to thwack

Parsons at the first opportunity, especially because he, like Mitchell, holds fast to the idea that all is politics; it takes precedence over truth and fairness. With Parsons dead and Walker with more than a passing interest in California, it'd be astonishing indeed if he wasn't trying to find more than one way to say that California was his turf and others could either stay away or get his prior blessing on what to do and say about important geographical matters in the state.

Dick Walker wastes no time letting us all know the score, who's in the know and who isn't about California agriculture. The first line of the review reads: "Don Mitchell errs when he says that as many as 200,000 migrant workers have been employed in California's agricultural fields; the best current estimate is more than 500,000, a truly astounding figure."[4]

What did Don Mitchell do to deserve this deadly heart shot right out of the gate (deadly because it's a number and one seemingly wildly in error)? The following is what Mitchell in fact wrote: "At times, as many as 200,000 workers have tramped up and down the state in search of agricultural work, passing through, making possible Parsons's 'color and geometry,' Starrs's 'pastoralism.' "[5]

Several things are worth noting about Walker's shot: Neither he nor Mitchell gives a source for the numbers. Neither mentions over what period of time these numbers apply, albeit *current* does not sound like 1942; Mitchell's book is confined to the period 1913 to 1942. Walker is talking about the number of people "employed," whereas Mitchell is referring to the number of people "tramp[ing] up and down the state in search of agricultural work," seemingly a big difference. Finally, from the context of Mitchell's use of the number 200,000 and the way he handles it, it doesn't seem to matter a whole lot whether the real number was 200,000, 300,000, or 104,551 people.

So I was wrong: Walker aimed for Mitchell's heart and gave the strong impression that he's a very good shot. But in fact, on closer examination, it would seem that he did little more than shoot Mitchell in a small toe.

Relative status more or less established with this opening salvo, Dick Walker doesn't want to lose Don Mitchell as a comrade in the never-ending war with nasty capitalists, and so in the second paragraph we're informed that Don Mitchell is way, way ahead of any labor historian who has dared to open nasty files on California's agriculture history. The reason that Mitchell is way ahead of any who came before him is that he "deploys a Foucauldian analysis of the panoptics of power in a way that cuts swiftly to the heart of labor relations in the fields."[6]

Although I've read Foucault, I have only the vaguest idea what, in this instance, constitutes a "Foucauldian analysis of the panoptics of power." And I'm not at all sure that Dick Walker does either; I think he's bluffing, making himself look smart by getting on the bandwagon and genuflecting to Foucault, an iterative habit of the times. Not only does Walker not give us one iota of a clue about what this knifelike kind of tool has provided by way of insight into Mitchell's work, but much of what he later says also strongly suggests that if Walker and Mitchell are into any kind of Foucauldian analysis, then it most certainly is not panoptic. (Forgetting Foucault for the moment, my *Unabridged Random House Dictionary of the English Language* informs me that *panoptics* means "permitting the viewing of all parts or elements; considering all parts or elements; all inclusive.")

Dick Walker is, in fact, talking out of both sides of his mouth at the same time on this issue of Foucauldian analysis in panoptic wrapping. A major and telling criticism of Mitchell's book by Walker (with which I heartily agree) is that Mitchell's view in fact is quite partial—*it is anything but panoptic.* Here are Walker's own words:

> Mitchell's focus on labor camps also comes up short as a representation of the landscape of California agriculture. The camps cannot *stand for the whole*; nor can harvest labor alone. Where are the crops and the elaborate field systems of the Great Valley? Where the irrigation ditches and waterworks? California agriculture constitutes a gigantic, integrated landscape of production *of many parts*. It is one vast territorial apparatus of production.[7]

Well, which is it? Surely, a panoptic analysis would not have been so blind or partial to have missed what Walker—smartly so—would not have overlooked? It sounds very much like Mitchell is giving us anything but a holistic or panoptic view in his historical narrative of California agriculture.

Dick Walker is less than a third of the way into his review before he's eager to tell us that Mitchell has called up "the ghost of Carl Sauer." We're told by Walker that Sauerians such as Peirce Lewis (he is no Sauerian by any stretch of the imagination) and Wilbur Zelinsky (neither is people-shy Wilbur, notwithstanding his Berkeley Ph.D. and Sauer associations) "come in for some hard knocks." Though Lewis and Zelinsky deserve plenty of hard knocks for their eyes-only take on ordinary landscapes (as I discuss at length in the essay "Arrogant Eyes"), they do not in fact get banged around much by Don Mitchell, and certainly not because Dick Walker has asserted as much. (Lewis probably gets included because of his well-known, long-time, close personal and professional ties to Jim Parsons.) Walker also tells us that "the late James Parsons is singled out for his well-intentioned but conservative effort to reclaim the landscape of California agribusiness from the social critics and restore the naturalized aura of agrarian tranquility."[8] Singled out, yes, but, as I've noted (in the preceding essay), unfairly so by Mitchell (and now by Walker) who allowed his "politics-is-everything" credo to override the need for honest and good scholarship.

Dick Walker is guilty here, at least as guilty as Mitchell, because he knows better, at least if he really knows as much as he claims to about California agriculture. Surely, he must have read all of Jim Parsons's pieces on California, and as I've made abundantly clear in my critique of Mitchell, Parsons was far from entertaining a simplistic picture of the "naturalized aura of agrarian tranquility."[9]

Now to a different matter, that issue that Dick Walker has "puzzled over . . . long and hard," because he considers himself "both an environmentalist and a Marxist." What he has puzzled over so long and hard are these questions: Does it or does it not make sense to continue being a reductionist? "Does the harnessing of nature or of human labor constitute *the fundamental site* of social power?" Or is there something more fundamental in the environmentalist position, one that focuses attention on "a vast array of manipulations and degradations of the earth and waters: drained wetlands, irrigation ditches, far-off dams, deep wells and pumps, pesticides"?[10]

All this hard thinking has led Dick Walker to the incontestably profound conclusion that "a unified theory of capitalist exploitation must be that production, or human work on this earth, is simultaneously a labor process and a natural process, mediated by technology (knowledge and instruments of labor). Control of labor is thus equally control of nature."[11] This kind of discovery that all of life is complex and can't be reduced to *primal* or *fundamental* causes is, generously put, sophomoric piffle.

Walker, notwithstanding his well-known embrace of Marxism, seems to in fact like a piecemeal worldview. For example, he accuses the Sauerian types, and others, of embracing a cultural geography that is *just* (and narrowly) cultural. But I don't know what he means, and neither would Jim Parsons, and he was about as Sauerian as they come. Parsons couldn't see a blade of exotic grass or a cow in a corn field without simultaneously seeing cultural, economic, political, and all kinds of other forces at play. Parsons, as I remember him, and as I have known all decent field-workers, didn't go around cutting up the world into economic geography, cultural geography, labor geography, environmental geography, and Marxist geography—just a few of Walker's many pigeon-nesting boxes.

Don Mitchell is all exercised about California's agricultural landscapes being invisible, and so, this review reveals, is Dick Walker. But what could they possible mean by this notion of "invisible" or "opaque" landscapes?[12] That California's agricultural landscapes are "visible" only to Mitchell, Walker, and other Marxist comrades who alone have the keen minds and sharp eyes to see that the state's agribusiness is prosperous in good part because of all the cheap labor?

Well, even assuming that Jim Parsons was blind in every sense of the word, and that I am too, and that every geographer who has ever given five minutes' thought to the state and its abundant produce is also blind as the proverbial bat, it doesn't matter one whit. All this breast thumping about "invisible landscapes" is just one more example of academic conceit, for if we combine all that geographers for all time have said and done with regard to agriculture in California, then the whole pile doesn't amount to a kitchen anthill in terms of improving the plight of poor Mexicans, Filipinos, or others who have given the state the distinction of having the most agriculturally productive piece of real estate the world has ever known.

All this nonsense about invisible landscapes not only is a timeworn academic conceit but, worse, is also elitist. There is nothing at all "invisible" about any of the state's agricultural fields to the millions of Mexicans who have worked for lousy wages and wound up with broken backs, damaged lungs, and shortened life spans. Nor is there anything at all invisible about the state's agricultural fields to the millions of sons, daughters, mothers, and fathers both in California and in Mexico who, at one and the same time, have lived with laboring kin and all of their ills, yet have benefited immensely in California and in Mexico from the awful wages and the execrable working conditions.

Why is it that for all the self-righteous railing that Marxists do on behalf of the world's downtrodden, the view is not from the field, but from the comfy offices of academia? Why is it that although capitalist exploitation is every bit as bad as most Marxists would have it in their sane moments, the ones with a heavy-handed take on California or Texas can't find their way

south and across the border, if only to put in context how rampant capitalism really is, and how heartily it is embraced, as well as to appreciate that
however bad Mexicans have been exploited in California, Texas, and the
Midwest, being able to work and get paid what they do in California is a
kind of heavenly gift compared to the utterly degrading life they have so
desperately fled?

Feminist Field-Worker Conceits

> The taste for emotion may become a dangerous taste; we should be
> very cautious how we attempt to squeeze out of human life more
> ecstasy and paroxysm than it can well afford.
>
> —S. Smith (1771–1845)

SIX articles by feminist geographers on fieldwork that appeared in a 1994
issue of the *Professional Geographer* deserve comment.[1]

One hard-to-avoid impression that one gets from these articles is that
there is something uniquely feminist and, shall we say, privileged about
certain kinds of fieldwork. Another is that feminists are somehow more
sensitive or attuned to the needs and emotions of the people they interview
or talk with in the field than are males or anyone not a feminist. Still another is that feminist geographers, or these seven at least (one article has
two authors), have an unusual relationship with the people they meet in
the field, one of "betweenness." Feminists (feminist geographers in this instance), in short, are different from the rest of us, especially males like myself who make no claims about being a feminist or anything else.

Cindi Katz is enamored with ethnographies; indeed, she wants to call
everything she does in the field ethnographic. For someone who has published so little with empirical substance (not unlike the other feminists in
this forum), this appropriation of a term common to anthropological research is odd to people who have a long tradition of spending a year or two
living with a people in order to produce a genuine ethnography (notwithstanding all the misplaced anxiety about text, authorship, and such). When
I think of ethnographies in geography, I think of nothing I've done, but rather
what people such as Eric Waddell and Barney Nietschmann accomplished
working in New Guinea and among the Miskito Indians, respectively.[2]

Katz's complaint that her applied work is marginal to what's happening in geography (by which I think she means it's hard to get published) is unconvincing; if the fieldwork she says she's doing is any good, and if it's really ethnography (doing more than occasionally talking to people in East Harlem as time permits, or living in the house of a wealthy family in a Sudanese village—her own admission), then there are plenty of highly respectable places to get published, and having done so she will surely get the academic kudos she seeks.

I rather suspect that a good long year or two living in a rat- and drug-infested tenement in East Harlem might well disabuse Katz and other feminists of another conceit, this fashionable idea of "betweenness" (a concept approvingly quoted by Katz as meaning "a position that is neither inside nor outside").[3] In all of my fieldwork, much of it probably more intensive than what Katz and the other feminists in this issue have done thus far, I've never thought of myself as other than an outsider. And how could I be so arrogant as to think that I was even in some in-between land? To be even half an insider, for example (referring to my work on wild horses in the West), I would have to be not just stealing horses and rounding up cattle in, say, Nevada's Long's Valley year after year, but also raising a family there, coping with the biting-cold winters, and finding ways to make a marginal living.

To work as an occasional advocate for this or that group hardly makes, by my reckoning, a field-worker an insider. None of us, feminist or otherwise, no matter how avowedly political, is more than an intruder, an interloper, an outsider. The truth is rather unflattering: we impose ourselves on people to get information to then be able to give conference reports and publish articles and books. Any academic serious about change and making a difference in a community will gladly give up her comfy and well-paid university position and take to the streets. Noise at conference podiums or macho writing in academic journals has virtually no effect whatsoever on what goes on in the cold, hard world of hunger, welfare checks, and men beating or raping women. For people who think otherwise, theirs is an unadulterated academic conceit.

Cindi Katz as feminist and self-styled seasoned ethnographic field-worker entertains other strange ideas. She states that "ethnographic and other forms of qualitative research [in geography] have had to conform to standards that are external to their constitution."[4] From all the evidence I have—my own highly qualitative research and the scores of papers, monographs, and books published by people who have earned Ph.D.s at Berkeley, UCLA, Chicago, and Syracuse (to mention but four geography departments)—this claim is patent nonsense. It is an insult to the integrity and firm convictions of scores of committed field-workers, among them a few real ethnographers. Katz, I'd surmise, has been reading too much Derek Gregory and first-generation David Harvey, people who have produced precious-few empirical insights into the real world.

Katz feeds us another liberal jewel that needs some reworking. She says that "virtually all ethnographers protect the anonymity of their participants."[5] On the assumption that within her liberal definition of *ethnography*, I am one of its producers, not only do I not protect the anonymity of my participants, but I also think it is quite wrong in a great many cases to do so. Not only is it important to know the source of charges and claims in order to be able to assess their worthiness, but crooks, exploiters, and people who rape the land should in fact be exposed. After all, such exposés are about as close as academics, like journalists, can come to making a difference in a world that is pretty bleak and sordid. I'd suggest that, whatever documentation would show about Katz's claim, she and other feminists who are so sensitive to exploitation and marginalized peoples change their modi operandi.

Equally off the mark is Kim England's rhetorical question about whether it's ethical to reveal the location of a study, in her case of Toronto's gays and lesbians. Does she really believe that a jargonistic article read by a small number of academics in an obscure journal (or even, let's say, that mighty flagship journal the *Annals*) that reveals the location of the study is going to provoke outing and physical violence?[6] Is it really true that gay bashers in Toronto wouldn't know where to find their prey without England's help? She and others who talk and write this way are taking themselves and their far-less-than earth-shaking findings—whatever they are—far too seriously.

These feminists might also pull back, or at least severely bracket, their embracement of "supplication."[7] It may make them feel different, and better, from aggressive field-workers like me, but, if my experience is any measure, their headstrong desire for hugs and kisses and leaving just the right empathetic impression often leads them down paths littered with deceptions, half-truths, and falsehoods.

I am not, of course, arguing that field-workers ought to treat the people they are involved with other than with respect and dignity. But if field-workers, feminist or otherwise, work from the premise that the people with whom they talk are again and again straightforward truth tellers, they are naïve indeed and need to be reschooled. In all probability, they have done little serious fieldwork.

This kind of misunderstanding of peoples—marginalized or not—and a lack of appropriate measures to discover truth take a different turn for England when she talks about her failed research among Toronto's lesbians. She says she had doubts about her lesbian research assistant; she felt angst over being a straight, white feminist; she didn't know what she could and could not reveal; she wondered if she was a voyeur (all field-workers are voyeurs); and then the whole project fell apart when a couple of telephone calls were not returned.

In my own fieldwork, blind alleys, refusals to return phone calls, and men and women alike who have refused to talk with me have been commonplace. Did England ever think of hanging out at lesbian coffeehouses and befriending lesbians? Of moving into the community? Straight or not, of getting involved in a lesbian outreach or hot-line program as a way of beginning to do, yes, an ethnography?

But then perhaps it's no wonder that England comes up empty-handed and that Katz and others can't get a real hearing for their "ethnographies." Even a cursory examination of their bibliographies and citations shows that, like so many other kinds of special-interest academics, these feminist geographers are long on polemic, self-pitying introspection and breast thumping and amazingly short on simple and straightforward descriptions of places and their peoples.

All of the above brings me to another conceit, by no means confined to this group of self-styled feminist geographers: the use of the off-putting term *Other*. These other people do in fact have identities: Imperial Valley farmers, illegal aliens named Jose and Marie who pick lettuce, midlevel executives who work for the Imperial Irrigation District—I could be a lot more specific. Why, then, this "Other"? It's as off-putting as the erstwhile and chic, "And meet my significant Other!"

Are badges and public declarations now required to be a legitimate feminist? Several of these feminist geographers seem intent on impressing us that they are "antiracist," not "heterosexist," and, well, . . .different. Does declaring that one is against racism or sexism place one on a higher plane, make one a churchly guardian of the rest of us thoughtless bumpkins who merely go about the grubby business of confronting racism and sexism eyeball-to-eyeball, with hot words if necessary? I'd be a lot more convinced if Audrey Kobayashi—the loudest shouter of the lot in this series of articles—had given us a single poignant example of how she personally took on someone who was racist.

In a somewhat related vein, Kobayashi makes much ado about the pervasiveness of essentialism and naturalism in the academy, and, I presume, in geography. This assertion is certainly provocative, and I suspect I'm not the only one who reads this journal who will think that her case is much overstated if not just dead wrong. Regardless, I'd love to see the hard evidence, and more than one or two citations. These kinds of categories may make for lively straw-man (oops! -woman) presentations to uncritical undergraduates, but as far as I can tell, they are hardly the stuff of a well-researched and heavily documented dissertation or book.

Kobayashi asserts that essentialism in any form is completely unacceptable. She has made up her mind, and nothing will convince her otherwise: "My position is that *all* social reality is constructed."[8] This fact will come as late if still great news to my wife who went through twelve hours of difficult labor to deliver our son on July 3, 1990. Had my wife known this eternal verity about reality, she would have been more than happy to ask me to carry our son for four and a half months, and to split the arduous labor time. This idea that all social reality is constructed will, of course, be an especially startling revelation to her because she just happens to be an evolutionary biologist who, to get my sperm and one of her eggs doing the right kind of dance, quite literally concocted her own chemical brew to shoot herself up with to get pregnant. As far as I can tell, this bit of imaginative biology was hardly a world in which *all* was social reality.

Finally, it would have been more than a little gratifying to see some concrete examples of just what it is that feminists can find out in the field that I and others like me cannot. Can they get inside, unlike men or nonfeminists, some of the mysteries of PMS? Do they relate to certain categories of men or women (and who are they?) in the field in such a way that they get information I cannot get—and *exactly* what is it that I cannot get access to? Do feminists approach—literally approach on foot—their subjects differently from nonfeminists? I'd honestly love to know, if only to try out, if not to appropriate, the methods. Nay, steal them and call them my own.

New Orleans Folks and Fictions

> Writers often complain about critics and criticism and yet they also complain when no criticism of their work appears.
>
> —Elizabeth Hardwick, *American Fictions*

HOW many different ways are there to portray a city? Among the ways of doing so, are there some ways that are clearly better—or, rather, more illuminating—than others, or are we to think of all ways as equally valid? Does it matter whether we genuinely pay attention to a city's diversity, or is it enough to characterize a city of a million or several million people in terms of rather broad generalities?

These and other questions came to mind when I returned to Peirce Lewis's *New Orleans: The Making of an Urban Landscape*.[1] It is perhaps the best known and most widely respected of a series of city "vignettes" written by geographers in the 1970s. Most of them were, I think it fair to say, if not dry disasters then forgettable.[2]

Lewis calls his book an essay, and he presents it as an answer to students whom, over the years, he had been telling that it was possible to recognize "the hard realities of urban life without treating the city as an economic machine," and that it was also possible to "draw a holistic picture of a place like New Orleans." So he spent some time in the city (an academic year is my guess; he really doesn't say how long) and with university geographers, planning-commission people, and even Mississippi River bar pilots. There are others to whom he is indebted for what he learned about this "big city in a short time—not perfectly, of course," and, as the last line of his acknowledgments would have it, he is "grateful beyond measure" to the "people of New Orleans—a fine folk."[3]

Following the bibliography, we're told that Peirce Lewis has, among other things, lived and traveled in most parts of the United States, as well as Europe and Asia; has studied "the political behavior of Negroes in urban Michigan, the culture of small towns"; and at the time of publication was writing a book on "the vernacular cultural landscape of the United States."[4] So, he is a student of people, a cultural geographer, a man afoot with eyes, mouth, ears, and pen and notebook on ordinary landscapes—we might assume.

Early on, in the first chapter, titled "The Eccentric City," Lewis says that "New Orleans is in a select company of American cities beloved by their residents and praised by visitors."[5] How he knows that statement is true, any more than Beltway pundits have a clue what's on the minds of ordinary folk in Missoula, Montana, is a complete mystery to me. As far as I can tell, it's merely a prejudice he brought to New Orleans before he arrived. As there's not the slightest indication that he talked about this issue to even a small sample of the city's inhabitants, to say nothing about a representative sample (awfully difficult to get for a city of 1 million people in the very best of circumstances, even by a professional pollster), there's no reason to believe it or many other kinds of citywide generalities he makes. They're about as valid as claiming that Poles are brash or brilliant, stupid or silly, aggressive or laid-back, alcoholics or auto mechanics. Take your pick. In truth—and it ought to be obvious—any such generalities can, at best, be nothing more than chockablock with exceptions, and so many of them in fact that such generalities are no better than the kind of twisted barnyard banalities and blather that are regularly voiced about "uppity Frenchmen" or "miserly Jews" or "aggressive fem working women."

The one thing we do know about any grand geographical grouping of several thousand people, to say nothing of more than a million (the population of New Orleans at the time Lewis wrote his book), is that there is very considerable diversity of opinions about, yes, almost *anything.* And even some modest display on paper of this diversity will be far more interesting and closer to the mark on the nature of a city than the most carefully imagined—to say nothing of written or spoken—generality about a whole city. This fact is why the very glib, globe-trotting "city synthesizer" Jan Morris is so frustratingly dead wrong about so much when she reaches for the one-, two-, or three-word phrase to get to the "essence" of Sydney, Rangoon, or San Francisco.[6]

In about the middle of this small book, Lewis talks about "the Latin American linkage," and comes up with a high figure of eighty thousand Latin Americans in New Orleans in the early 1970s.[7] There was, he says, once a large Cuban population, and there are now (at the time of his writing) substantial numbers of Hondurans, Guatemalans, and Nicaraguans. But what do we learn about any of these Central Americans or Cubans? Do we hear anything about where they live, why they have come to New Orleans, whether they like the city, whether they want to return to their native country, what they eat, how they are perceived by non-Latins, whether they wax romantic about the city's history, or whether they can even spell the name of that great river that may be a block or two away from where they live? Nothing. Not a word. Not a word for one of these groups, not a word from a single one of the individuals who might be—heaven forbid—a "representative" speaking for the whole of the group: Hondurans, Nicaraguans, Cubans, and so on.

Lewis isn't interested in any of these voices—not a one. In the whole of this book we hear not the single voice of a rich or a poor black, a Hispanic, a Cajun, a woman, a child, someone new to the city, someone born in the city, or . . . ad nauseam. We get not a single hint that Lewis even talked to half a dozen of any or all of these people. So, then, how could Peirce Lewis "know" how the city's residents "see" their city, how they "feel" about it, or whether they love it or hate it, only tolerate it, or don't give a flying anything about it? Pure and simple, he can't. He just does not know.

We don't get from Peirce Lewis a picture of any variety at all; what we get is a very selective exercise in mining some historical documents, a lot of maps that are far too gross to reveal much of anything (if you want to be geographic, then just put in a lot of maps, no matter how irrelevant or unrevealing), and about twenty photos, with a heavy emphasis on architecture.

I don't have any idea how much Peirce Lewis drove around New Orleans and got out of his car (other than to take the photos—and he may not have gotten out of the car for even this reason), but it appears that just about every line of the book could have been written not only without leaving his apartment or the libraries he used, but even without (the photos he took excepted) going to the city. My guess is that Peirce Lewis could have stayed in his snug office at Penn State and had interlibrary loan send him the books and maps he needed, and the book he wrote wouldn't have been much different from the one he wrote by living there for part or all of a year.

None of the preceding is to say that Lewis hasn't produced *something* of value, but what that something is has to be kept in perspective. He's given us a nice, brief, and cleanly written historical geography, I suppose. But—again—there are no voices, no smells, no sense of the look and feel of a single street, a single neighborhood, the ordinary grinding and very quotidian lives of all those interesting people that *are* New Orleans. To be sure, the city is also its history, its geography, its buildings. But what makes it a whole lot more, and gives some real meaning to the word *holistic* (what Lewis says he's after), are all of its people and their stories: short ones, long ones, sad ones, happy ones, as different and unique and fascinating as each individual is in his or her own way.

In an important sense, I'm talking about the kind of book I would want to read on New Orleans, and the kind of book I would have written.[8] So, from this vantage point, Lewis can't be faulted. But, then, having said it, I still must return to Lewis and his heartfelt thanks to all those "fine folk" that he mentions, to this city that he claims to have "learned" (his word) so much about, and to this very "romantic" place where the "ordinary Orleanian *assumes* that everyone else shares his admiration for his native city." To all of it, I say, Hogwash. Peirce Lewis doesn't know that the people who live in New Orleans are fine folk anymore than he knows that people who live in Harlem or Ann Arbor are "bad folk" or "kinky folk" or "very intelligent folk." He doesn't know because he didn't talk to enough people outside his library and university circles to stick in his back pocket. And although I'm prepared to believe that Peirce Lewis may well have a love affair with New Orleans, the reasons, I am sure, have little to do with having genuinely "learned" the city. He didn't meet the challenge he set before his students. And he's right in noting in the acknowledgments that, with regard to his success or lack thereof, he's "too fond of New Orleans to be very dispassionate."[9] It's too bad that his passion for the city didn't translate into a serious encounter with all the people who make the city of New Orleans whatever it is.

Arrogant Eyes

When men are most sure and arrogant they are commonly most
mistaken, giving views to passion without that proper deliberation
which alone can secure them from the grossest absurdities.

—David Hume

A fictional scenario. On the balmy night of April 1, 1998, four geographers
of academic renown sat side by side on a cement parapet across the street
from the Havana Libre Hotel.[1] Their names were Elihu Gray, Eliphalet
Black, Evander Green, and Zephaniah Brown. For the previous fifteen
nights, they had arrived at this same spot at nine in the evening, and they
stayed until two in the morning, at which time they repaired to their $120
rooms on the eleventh floor of the Havana Libre.

They came to the parapet outside the hotel each night because they had
collectively decided to write a definitive essay titled "The Havana Libre:
Sight and Street Symbol of Cuba in the Tourist Years." It was, in the tradition in which they had become famous, an exercise in describing what they
liked to refer to as ordinary landscapes. This kind of description was based
on "seeing," for with minor exceptions they held firm to the belief that seeing represents a special way of knowing, superior in all important respects
to other ways of knowing the world. Indeed, they were convinced that their
trained eyes were special, so special in fact that what they saw was the
measure of reality. What people have to say about who they are and the
place where they live are at best a secondary kind of reality, unless such
local knowledge has found its way into the historical record. It must be
noted that these men loved books and libraries, and they firmly believed
that whatever they had seen with their naked eyes had a history and had to
be understood principally in historical terms.

During the day when these four geographers were not sleeping, eating, or wandering through Havana's famous Lenin Cemetery—which, it should be noted, held considerable interest for all of them (though why, with the exception of Zephaniah Brown who had an unerring fascination with death and the geography of cemeteries, is most certainly an unfathomable mystery)—they poured through old issues of *Granma*, the state newspaper, and looked at yellow manuscripts and books with broken spines in the Jose Martí National Library that fronts on the Plaza de la Revolución. There they were able to discover, among other things, when the Havana Libre had been built, how many rooms it had before the revolution, exactly where the poker and baccarat tables were in the casinos in the Batista era, and what kind of building materials had been used in construction. (The provenance of the limestone used for flooring in the kitchens and bathrooms was of unusual concern to all of them.) They had also come across records showing the percentages of the hotel (when it was the Havana Hilton) owned by Mafia families in Miami and New York. And they even found out which rooms Fidel Castro used for work and which he used for bacchic evenings with his many mistresses in the first five months of 1959 after he came to power.

Evander Green occasionally got restless, and on some nights he insisted on sitting with taxi drivers in front of the Hotel Inglaterra and the Hotel Nacional. He had long had a penchant for observing and recording life while sitting in moving and stationary cars, or atop his trusty high-bar Harley. He wasn't much interested in talking with the taxi drivers, and in fact he couldn't; he spoke only the most rudimentary Spanish. He had in mind that he would like to incorporate into the descriptive essay on the Havana Libre some of what he saw from afar at these other hotels. He believed that composite descriptions passed off as descriptions of a single landscape were quite acceptable, a new and novel form of finding similarity in diversity, he thought. A little-known fact about Evander was that a literary hero of his was Gail Sheehy, who had once written a rather famous little book based on a composite hooker known as Red Pants.[2]

Zephaniah Brown would also occasionally get restless, and on one or two occasions when some in his presence began to lean on one another and snore from boredom, he would slip into the disco not far from the parapet where they all sat. There, while doing intensive fieldwork on the disco snack menu and the red-and-blue neon signs advertising Cuban beers, he was surrounded by hustling *jineteras* (prostitutes). As Zephaniah took copious notes on the menu and the signs, *jineteras* made eyes at him, and now and again one would approach and ask if he wanted to dance or, better yet, take her to his room. No, no, no! Zephaniah would insist, without raising his head or slowing his note taking, saying no more than he did because he was unable to understand what they were saying to him.

On this particular night of April 1, 1998, just a couple of minutes before midnight, a somewhat unusual event took place: all four pairs of eyes simultaneously caught sight of a tall white American whom they immediately recognized. None of these geographers in fact knew that much about this person, though each agreed that he was, by all measures, a disciplinary black sheep. In company with Owen Lattimore and Bill Bunge, he had the ignominious distinction of having been blackballed for life from the very discipline these gentlemen of weighty moral distinction called home. When they caught sight of him, they could not resist describing what they saw, thereby creating truths as solid as any for which their stellar discipline proudly takes credit.

I see that he's wearing a dark-red baseball cap with the letter *C* on it.

Indeed he is. Not only is he a baseball fan, but because that cap represents the University of South Carolina, he either lived in the state sometime or went to school there.

Or he might well have a brother or sister in school there.

Notice that he's holding the hand of a young Cuban girl.

A *mulatta*, it should be noted.

I understand that he's been married several times, and given his well-known reputation for, shall we say, licentious behavior, it's fair to conclude that he's undoubtedly got something going with this young Cuban we now see him with.

Because she looks so young for him, maybe she's the daughter of a Cuban woman he's involved with.

Or could he be, yes, involved with her?

No doubt.

Which?

No doubt we shall know shortly!

But wait. Knowing him and his reputation, and what we've read about these young girls called *jineteras*, or street jockeys, I would not be at all surprised that he's found himself an obliging whore for the night and they're headed for his hotel room.

Of course, it must be so. After all, she is dressed rather well, not unlike others we are seeing here on the street who also seem solicitous.

Have you noticed the gold chains around her neck? And that silk dress and high heels? She undoubtedly comes from a monied family, one that escaped Fidel's clutches.

But then . . .

Wait! Did you see her just put three fingers inside his pants? Surely, it indicates . . . well, what does it indicate?

I would imagine that they are quite intimate, hardly unfamiliar with each other.

Hmm . . .

At the corner of the Havana Libre, or more properly the now-dark gift shop, at the stroke of midnight, the black sheep and the young Cuban girl whose hand he's held since they came into view turned the corner and disappeared from sight. In the dark shadows and out of view of the police, he let go of her hand, at which point she said to him, as she had twice since she'd encountered him, *No me vas a llevar contigo?* Once again he responded as he had previously, *No tengo interes en lo que quieres hacer.* She frowned, and he gave her a parting kiss on the cheek, and he said, *Que este bien.* She walked away and disappeared into the darkness.

At two o'clock (it was actually 2:02), the four geographers got up from their cement seats, weaved across the street among honking taxis, *jineteras*, and several hard-hustling *chulos*, and went up the seven steps to the cement walk that led to the front doors of the hotel. Just inside the lobby, Zephaniah Brown said that he would be going directly to his room. Despite his interest in this fascinating Cuban scene, he was eager to update his 1969 S.M.A. and county maps of "Lutheran Bodies" and "Presbyterian Bodies," for he was sure that in the Upper Midwest, three and two data points, respectively, had to be subtracted from each of the maps to bring them up-to-date. But where, he had wondered for days—nay, weeks—would he squeeze them in among an already mind-numbing array of "population boxes" that cluttered the maps at a scale that he had never bothered to specify? It would be a difficult decision indeed, one that would keep him awake until sunup.

Eliphalet Black said that he also had some urgent work to do. He was in the process of going back over old and moldy notes, looking for documentation to an earlier book of his.

Evander Green had other matters on his mind: whether he had lost money on a recent sale of Microsoft stock. He wanted to fax his stockbroker straightaway, which was actually possible to do in this hotel, such was the Castro government's concern for tourists and their dollars, and its own blatant if unacknowledged turn toward capitalism.

Elihu Gray was in a different state of mind. He wanted to have a drink or two in the first-floor bar before going to bed. He had worked his eyes to exhaustion, and, he had to admit to himself, he was more than mildly titillated by what he had concluded about this disciplinary black sheep who had come into view in the closing minutes of April 1. He wanted to make some notes before he forgot all that he had seen.

Imagine, then, Elihu's surprise when he got to the bar, and who should he see sitting by himself and apparently arguing with the bartender, but the black sheep himself! Elihu took a stool three removed from where Korski sat, and without acknowledging him—not even sure that he would be recognized—he tried to listen in on the conversation. But it proved nearly impossible, for Elihu spoke and understood less than a dozen words of Spanish, about the same number as his longtime Northeast State University colleague Zephaniah Brown. Still, Elihu thought he overhead the two of them arguing about the U.S. embargo of Cuba, and he thought he heard Korski say that he was in favor of the embargo. (In fact, Korski and the bartender, good friends in spite of the fact that one loved and one hated Fidel, were arguing about which had been the best of the ten cocks they had watched in a four-hour series of cockfights in the coastal city of Trinidad two days earlier.)

Elihu was into his third Cuba libre when Korski let go with a cackling laugh and then broke away from the bartender and took his drink to a small, circular mahogany table surrounded by four wicker chairs. Elihu downed his drink and ordered another one, and with this drink now in hand, he got up the unprecedented courage to go over and ask Korski if he could join him. The rum had loosened him up to the point that he felt absolutely daring, emboldened like he could never remember. Elihu had some serious questions he now wanted to pose to a person he didn't like very much. In fact, he didn't like Korski at all, and the principal reason was that many years earlier, Korski had had the unprecedented impudence to write that scathing critique of Elihu's bosom buddy Eliphalet Black. (Remember, my questioning postmodern reader to whom I owe a personal word, how well that went over, as Korski was a mere assistant professor and the Don—known as the Cardinal among some of his Eastern State University colleagues—was a full professor of considerable distinction!)

The introductions over, Elihu, now energized by all that rum coursing through his veins and wondering why it had taken him so long to discover the incomparable purity of fifteen-year-old Cuban rum, said, I saw you tonight with your Cuban girlfriend. She *is* a quite attractive *mulatta*. And young, too, I must say.

A beauty indeed, but I must confess that she's not my girlfriend. Incidentally, she considers herself *trigueña*. I don't think she'd be at all flattered if you described her as a *mulatta*. These distinctions are self-ascribed, have racial overtones, and are personally important.

Well, then . . . your wife, if I might say so with congratulations!

Korski laughed. I don't even know her name.

Elihu giggled, self-consciously. He was just plain embarrassed. Now he said, Ah, then you, no different from all those nineteenth-century French artists and novelists, just cannot stay away from women of ill repute!

How perceptive of you.

Elihu smiled, and he moved his chair closer, then said: Ummmm . . . If you really don't mind another of my intrusions, I must ask, why was it that you were holding her hand? *That* was the great giveaway that led to all my—forgive me—probing questions.

He laughed without restraint, and he thought, Is this guy ever uptight. Then his knee banged the short table, and his drink flew into one of the empty chairs. Oh shit! he exclaimed. He looked in the direction of the bar and said (affectionately, it should be noted), *Coño! Otra, por favor.* He then turned to Elihu and said, There's a simple explanation about what you were observing. As I came up La Rampa and just before turning the corner to pass in front of the Havana Libre, this gorgeous *chica* that you've called a *mulatta* asked me if I'd hold her hand. I had, up to that moment, never seen her before. Because she would be passing in front of all those police, by holding my hand she could avoid being picked up on a prostitution charge and winding up in prison. The police, as you may know, will never touch a *jinetera*, no matter how well known, as long as she is with a foreigner. Anybody will do just fine. I was convenient. And I'm always obliging. He smiled.

Well . . . uh . . . that gesture then of hers—if you don't mind me asking—putting those fingers inside your pants. What was *that* all about?

Sticking it to a cop she detested. A *caoba sin cajones,* she said. From time to time, he apparently hits her up for money to work La Rampa.

Oh, I see, I see. . . .

Any more questions?

I didn't know you were a great baseball fan and had emotional ties to South Carolina.

I hate baseball, the most boring game imaginable. And I've never been to South Carolina.

Well, then, if you don't mind me asking yet one more question: why do you wear that Gamecock hat?

My son's name is Cole.

Ummm . . . yes—waiter! When the waiter didn't react to a word pronounced perfectly in English, Elihu Gray turned to Korski and said, How do you say . . . is it *coño?* Korski nodded. *Coño!* he now yelled at the waiter.

In the summer of 1968, I spent six weeks in Costa Rica on a field course led by Jim Parsons and other Berkeley luminaries. Carl Sauer, then an old man but still of sound mind, joined us for several days. On one of our early-morning trips to see what we could see from the roadside, I and a few other students sat with Sauer on the side of a mountain and listened while he offered an explanation for how he came to the conclusion that the field in the foreground had been cleared and burned within the past twenty to twenty-five years. Sauer did not encourage any of us that morning to wander off and find the farmer who owned or cultivated that piece of land and see whether his visual guess was in the ballpark. But I and another student, whose Spanish was better than my own, did wander off. From talking to a couple of farmers, we learned that Sauer was off by a good fifty years.

Twenty-nine years and nine months later, in April 1998, I found myself in Cuba. Over the next eighteen months, the island and its politics would become something of an obsession. I initially went to Cuba with the aim of doing little more than taking a lot of photographs (which I did—four thousand shots to date). But within two hours of my arrival, all of my best and worst snooping instincts went into high gear, and from that point on, I could not get enough exposure to Cubans—ordinary Cubans on the street, in their homes, late at night in bars, anyplace I could find them. My preoccupations were many, but they always returned to matters political: feelings about Fidel, the accomplishments of the revolution, their attitude toward the American embargo, how they felt about Americans, and what the whole *jinetera* system was all about, as it was so central to the tourist dollars pouring into and shoring up the sinking island. I have now traveled the whole of the island and spoken to more than five hundred Cubans. All this explanation I mention as prelude to a single example, relevant to what I have noted about Sauer and to what will follow.

One of the frequent claims of foreigners (invariably Europeans) I have met in Cuba is that Cubans hate Americans. In all my travels and encounters in Cuba, I have thus far met exactly one Cuban who expressed this sentiment to me in any form, and it was a *jinetera* in distant Baracoa who was thoroughly pissed off at Clinton and his "imperialistic ambitions." She did not, from what I could tell, dislike me, or Americans as such. Other than this one example, I found again and again that Cubans embrace Americans. Americans as people are seen as good people in Cubans' minds, and as something quite apart from what Cubans feel about American foreign policy or the embargo. To test the proposition that I was not merely hearing what Cubans wanted me to hear about how much they loved me as an American, I would, from time to time, tell people I met that I was either Australian or Canadian. The reaction toward Americans was, I would discover, no different.

How, then, could I have been hearing so much—to me—obvious nonsense from Europeans? One reason was that in virtually every case I discovered that my European know-it-all didn't speak more than a few words of Spanish. The second thing I invariably found was that one purported instance of one Cuban not liking Americans was, to my European informants, tantamount to all Cubans not liking Americans. And the third thing I ascertained was that with rare exception, these Europeans had some fundamental beef with U.S. foreign policy: regarding Kosovo, global cultural influence, and most especially the "immoral" embargo.

With all these thoughts as preface, it is now worth turning to Peirce Lewis and an article in the *Geographical Review* titled "The Monument and the Bungalow."[3] In the article, Lewis is keen to make two key points: students must learn to see commonplace things that ordinary Americans don't see or ignore, and they need to master a vocabulary for seeing. It is, to Peirce Lewis, *all* about seeing. Nowhere in this article does he so much as once mention the value of talking with people who own, live, breathe, and daily change the ordinary landscape that they inhabit. In other words, the value of asking what a landscape means to locals, as opposed to what it may have meant a hundred years ago, or may have meant to architects, historians, and builders who built some houses and some monuments and put up some plaques to dead soldiers in wars long forgotten in the common mind, is nil.

It is simply extraordinary, not because students can't be taught to give architectural names to this and that thing that they hadn't given names to before and to even understand what *might* have been some of the history behind putting up certain buildings or monuments, but because an awful lot of this kind of information means absolutely nothing to the vast majority of people in any town in America, Costa Rica, Cuba . . . name the place.

Ordinary people don't "see" history of the sort that interests Peirce

Lewis. They see a small-town Pennsylvania monument as a place where they enjoyed their first joint, or got their first kiss, or took Mom and Dad for a group photo the first time they came to town. These things are the kinds they care about, and Peirce in his maniacal way of reducing reality to seeing can't possibly know any of it because the last thing he wants to do is talk to local people about what monuments, California bungalows, or farmers' barns mean to them—if they mean anything at all.

One way to appreciate how different Peirce Lewis's reality-known-through-seeing is from reality-known-through-talking (what I am obviously advocating) is to look at what kinds of generalizations and conclusions he would have come up with had he gone to San Pedro Sula, Honduras, and asked this rather general question: What happened to food and clothing sent for the refugees of Hurricane Mitch? This question, broadly, is what I asked in an article appearing in the very issue of the *Geographical Review* in which Peirce's article was featured.[4]

I found that refugees living in *albergues* (shelters) were not getting any of the food sent from abroad. I did not *observe* this fact, for even if—let us say—all food sent from the United States had come in packages clearly marked as such, I might not have seen the labeled containers or they might have been destroyed before my arrival. Furthermore, there was nothing about the appearance of either the adults or the children that could have told me what kind of food they were eating or where the food had come from. Thus, without belaboring the point, I simply cannot see how I could have concluded what I did without *asking* directly—*asking* people in the shelters (and in more than one shelter) and *asking* the ones who were in charge of distributing overseas donations.

My guess is that based on observation alone, no matter how well trained the eyes, Peirce Lewis—whom I assume is a man of good intentions and not eager to attribute bad motives to others—would have concluded that because the people in the shelters were obviously not starving, nor were they walking around naked, then they were probably receiving the aid that had been sent for them. Even if Peirce had been brave enough to ask direct questions of one or two people somewhere in the pool of the ones who would have answers to such questions, this very small number might have done little more for Peirce than make him mildly skeptical—his degree of skepticism depending, of course, on how credible and strong the answers seemed at the moment and his ability to judge them as such, and on his initial assumptions about the honesty of Hondurans.

For a second example, we can imagine how Peirce might have reacted to the food and clothing that were in warehouses controlled by the San Pedro Sula Chamber of Commerce, which, as I found, was just sitting there or being ripped off and not going to the refugees. A purely visual inspection inside the warehouses might well have led to the easy—and even reasonable—conclusion that, based on the number of people who were busily moving goods about, and seemingly doing something, food and clothing were surely being distributed to refugees. That a warehouse looked full might have suggested to Peirce's eyes that at least in San Pedro Sula, the food and clothing were moving quickly from the docks to the warehouses. He might have been further reassured by reports he had read in newspapers that Americans, to say nothing of Mexicans and Europeans, were making very large donations for the victims. Yet only by asking, and being rather aggressive about it, would Peirce have discovered, as I did, that what was in the warehouses was, with the exception of what was being ripped off, there from the beginning. In other words, little or nothing had been distributed. What I learned was that everything was "being sorted and classified," and the time had not yet arrived—and no one knew when it would—when enough sorting and classifying had been done such that it was time to distribute food and clothes to the refugees.

A number of other claims in my article can be subjected to more or less the same logic. In the last analysis, then, had Peirce been true to his dictum that seeing-is-reality, and usually reality enough, his overall conclusion would have been that the relief effort, on both the giving and the receiving ends, had been a booming success. Furthermore, he might have then argued that what he found was new and refreshing: Honduras, his narrative would read, is a shining example of "good behavior" in the relief arena among Third World countries. I, alas, came to a quite different conclusion.

Peirce Lewis and his ilk might well respond that they simply ask different questions, and *had* he, for whatever reason, been put in a situation to ask the kind or kinds of questions I did, then he most certainly would have behaved not much differently from the way I did. He would have subordinated the findings of his eyes to what he heard through a variety of informants, and in his own way he would have been just as aggressive as I was in seeking answers.

This defense seems, at first blush, quite reasonable. However, the problem with this response is that Peirce Lewis, no different from anyone, is a victim of his own philosophy (just as surely as I am). That is, because he has elevated seeing-as-reality to such a lofty plane, it is rather unlikely that he is suddenly going to shift gears and stick his eyes in his back pocket and let his mouth, and other mouths, create the reality he wishes to narrate. Furthermore, as far as I am aware, there is precious little in his published research that shows more than the most minimal attention to mixing and jostling and hearing the common voice about ordinary landscapes. Peirce—as he has told me in person—would say that anyone need look only at his "impressive" effort on New Orleans (see the previous essay) and his article titled "Small Town in Pennsylvania" to appreciate that he is no different from me. But even a cursory examination of this small book and this article to which he often returns will show that his narratives betray his claim. For example, in "Small Town in Pennsylvania," Peirce claimed that he had talked "with hundreds of its citizens in innumerable situations" over a period of some fifteen years.[5] Yet there is nary a hint of such revealing conversations in the essay. Indeed, Peirce devoted virtually all of his effort to the history of Bellefonte, or his attempt to suggest—but not to prove—that the town is "not atypical of its class." When he was not summarizing or distilling what others said, he was talking about what he saw—not what he heard or learned from all he had presumably come to know through long residence in nearby State College.

Furthermore, and Peirce would be the first to note it, there is the real problem that you do not suddenly get a "good nose" for the right questions, or know how much to push an informant, or suddenly acquire the guts to ask uncomfortable questions. The very kind of in-field training and experience that Peirce claims is so necessary for "seeing" landscapes in new or interesting ways takes time. And so it does with learning how and when to ask questions that matter.

There is yet another issue: Just how *well* does Peirce see if he's not willing to complement his seeing, in a major way, with a lot of gritty mixing, snooping, and chatting? My guess is that he often sees a lot less than I do, and it is not because I know all that much about the people, places, or landscapes that, sometimes on a whim, I wander off to in search of some adventure and new—if only for me—knowledge. What I consistently find is—no different from anthropologists—that the interesting categories, questions, and issues are on the minds of the people who define and live the realities I choose to enter as a foreign intruder. (My biologist wife of more than twenty-five years, very much the scientist and ever attuned to theory and hypotheses, invariably asks me as I'm walking out the door on my way to some new place and unknown adventure what I'm looking for. And my stock and dead-serious answer has always been: Gorgeous, I haven't got a clue. Ask me when I get back.)

There are several things to note about Lewis's philosophy that—to me and others who have done a little dirty fieldwork—are disturbing (and I won't belabor these points because anthropologists have beaten this ground to death). For openers, Peirce's reality-through-seeing is simply elitist. Because the people who are daily surrounded by the monuments and live in the bungalows of which he writes so confidently are not asked how *they* see them, or what *they* see in them, the oral or written descriptions produced by Peirce are his highbrow academic take on someone else's reality. (It is true that a variant of this argument can be leveled against me, and with reason—but this issue is best left for another or longer essay. For example, how else can I describe Cuba in the nineties except in my own language—which means my own personal history—and, when discussing Fidel's concept of freedom, do so in terms of my own concepts of freedom, which are so radically different from Castro's that I still shake my head when I hear him use Abe Lincoln as his model for freedom?) Because the ordinary landscapes that Peirce is eager to describe are not his—he's not describing the neighborhood where he lives or his university environment—it seems only reasonable that we should know and understand not only what the inhabitants of these landscapes see or don't see but also that such interior or inside views have, yes, priority over Peirce's cute or fetching academic descriptions. And they have priority irrespective of how commonplace, mundane, or unremarkable what locals see or say about what they see is; it is simply elitist to either claim or imply that it's important for the inhabitants of Bellefonte to know that there is nary an Italian name among the four thousand on bronze plaques on the War Memorial. And it doesn't matter one whit more because a considerable portion of Pennsylvania's contemporary population is of Italian ancestry (which I'd bet most Italians in Pennsylvania don't know). To believe otherwise is to be blissfully ignorant of how the hoi polloi live in, think about, and see the ordinary landscapes that monopolize the minds of Peirce Lewis and geographers like him—D. W. Meinig, Wilbur Zelinsky, and a whole lot of other elitist academics— most of whom haven't got a clue that their small world is an awfully long way from all those Italian hard hats all over Pennsylvania who would say to Peirce's concern about the War Memorial plaques: Man, who gives a damn? Let me tell you what's really important.

• • •

I telescoped in on Peirce Lewis and a recent article by him and one by me, and some of the problems that his worldview presents. It is, I think, useful to broaden the view and look at the matter from the standpoint of how he and others (Jackson, Meinig, and others) have *talked* about what they do. Admittedly, how we describe or talk about what we do or have done is not always the same as what we do, but in the absence of other evidence, or as a way of making a case more secure, it is certainly helpful to look at what has been said and what it seems to imply. In what follows, I am cribbing extensively from a paper I wrote, but never published, in 1988.

J. B. Jackson has noted: "I have long been a strong advocate of learning about landscapes from firsthand experience: by looking at it, traveling through it, even living in it, however briefly."[6] How much Jackson may have talked with locals about the landscapes about which he wrote, I do not know, but it is not at all clear that he did very much talking. The films about him late in life, though showing him mixing and chatting with locals, don't, for me, make the case.

D. W. Meinig, who is rarely left out of any kind of discussion about ordinary landscapes, is very much like Lewis when it comes to the priority of the eyes over the mouth. He has written: "Landscape is defined by our vision and interpreted by our minds. Strictly speaking, we are never in it, it lies before our eyes and it becomes real only as we become conscious of it."[7] In a plea for an appreciation of localities a number of years ago, Meinig called for a "special education," one that would involve fieldwork, one that would produce studies of landscapes and places that get away from "wooden stereotyped descriptions of landscape elements with rather simplistic interpretations."[8] Yet, to Meinig, fieldwork involved only one's "eyes, ears, nose [and] touch"; the mouth—the ability to talk—is not considered important enough to mention.[9] Thus, to judge by what he has written, Meinig, and presumably the people who see him as a role model, is content to ask people at the "chamber of commerce, [the] public library, [the] historical society, or [the] local university for readily accessible descriptions of what the people of the local community are like, how they live, and interact as social groups."[10] For Meinig, knowing landscapes comes about through reading, what one finds in books, articles, pamphlets, and memoirs.

For anyone who knows anything about D. W. Meinig or how he has done fieldwork, all of what he says fits perfectly with practice. When he wrote his short book *Imperial Texas*, his idea of getting to know Texans and

Texas was to jump in a Hertz rental car and drive around the state and stay in nice motels, for a week or so. No doubt, Meinig was able to cover a lot of ground and get a great sense of regional differences in that rather small, homogeneous state.

It's worth returning to Peirce Lewis and some of the seven axioms he has identified for reading the landscape.[11] (I would hate to guess how many undergraduate and graduate students across America have been fed these "indisputable truths" by geographers not much different from Lewis.) He notes that landscapes are hard to study by "conventional academic means." He draws attention to the need to read fugitive and uncommon material. His seventh axiom states that "most objects in the landscape—although they convey all kinds of 'messages'—do not convey those messages in any obvious way." Lewis, like Meinig, turns to books, and to seeing to know landscapes. "To be sure," he writes, "neither looking by itself, nor reading by itself, is likely to give us very satisfactory answers to the basic cultural questions that landscape poses. But the alternation of looking, and reading, and thinking, and then looking and reading again, can yield remarkable results." Elsewhere, Peirce says that the "rewards can be greatly multiplied if one draws pictures of what one sees."[12]

A telling—yet not surprising—bias of Peirce Lewis, D. W. Meinig, and J. B. Jackson is that all of them believe that material culture has priority over breathing humans, those poor souls who just happen to be using the material culture that is of academic interest. In a telltale article titled "The Beholding Eye: Ten Versions of the Same Scene," Meinig identifies ten ways of seeing landscapes—as nature, as habitat, as artifact, as system, and so on. Nowhere did he speak of "landscapes as people." Elsewhere—for example, in *Southwest: Three Peoples in Geographical Change, 1600–1700*— Meinig's people are flat, disembodied, invariably a member of some rather vague group, as if history were made by faceless committees. Meinig is content to construct contemporary "culture regions" of landscapes he comes to know only through historical research. That regional boundaries bear more than a passing resemblance to the ones known to locals is a hypothesis to be tested, not a truth that arises of necessity through library research or suggestions made concrete and enduring by doyens in a discipline putting brush-stroke lines on a map.[13]

J. B. Jackson has claimed that the "individual dwelling" is the elemental unit in the landscape. In his words, it is "the oldest and by far the most significant" element in the landscape, because, he has noted, "first comes the house, the most reliable indication of man's essential identity."[14] But is this belief not misplaced emphasis by Jackson and by Meinig and others who give priority to the material landscape—in whatever form? Do they not have matters backward, if not inside out? It would seem to be a simple matter of cause and effect, of agent and product. Material culture, the visible ordinary landscape, would not be there or, were it there, would not be changing as it always is were it not for people, individuals. Furthermore, a primary emphasis on the built environment encourages a nondynamic view of landscapes. The built environment does not change its meaning; its meaning changes because local users are changing—in mind, in purpose, in the way they are using a landscape. The individual, then, is prior to, more important than, his material constructions. It would be obvious were one not strapped with the blinding whole-hog premise that seeing-is-reality.

J. B. Jackson's misbegotten priorities led him to the claim that a city is an "entity," that "much writing on the American city tells of its fragmentation, which comes from our predilection to see fragmentation." Jackson would have us see cities much as biologists view ecosystems: as systems.[15] In *The Public Landscape,* Jackson wrote: "The highway stands for unity. It is that installation which joins one part of the public landscape to another, which enables organized society to make its influence felt everywhere." Meinig would seem to agree with this proposition of wholeness. In an effort to distill Jackson's philosophy, Meinig has said: "Landscape is a unity, a wholeness, an integration, of community and environment."[16]

The proposition that landscapes are interrelated systems of parts and processes is a comforting scientific view, and it is one that gained ascendancy in the social sciences through the 1960s fad of systems theory and systems analysis and, later, various turns on Marxism. But the assumption of wholeness, attractive and even necessary as it may be for spinning abstract theories about society, does not square well with reality. The fragmentation and disjunction that others see in cities—but Jackson does not—are not merely matters of opinion, of using different lenses, of long learning and perspicacity. Just because elements in a landscape abut or people walk the same streets or drive the same highways does not a system make—except in a rather trivial sense. It would, for example, be a challenge of some note to demonstrate that down-and-out African Americans in south-central Los Angeles have much to do with the quotidian lives of the overwhelming majority of the city's middle—or upper-income whites—or vice versa. That members of one group are aware of members of the other group, that they see one another here and there on streets and through car windows, that high-class whites vote on welfare issues that affect these people and may even make occasional purchases from ghettoized blacks or Hispanics (such as illegal drugs), does not make a case for wholeness or unity. By this reasoning, virtually anybody could be characterized as being part of hundreds or thousands of "landscape systems"—a proposition that is, if not vacuous, surely of little value. But again, to the extent that there are indeed interesting and nontrivial interrelationships in any ordinary landscape, this matter is to be demonstrated. At best, wholeness is a hypothesis to be made operational and tested through real fieldwork, by letting locals speak, by coming to know what they know, feel, and do in their ordinary lives. In this light, wholeness is relegated to a different and less exalted category within an epistemological scheme concerning knowledge of ordinary landscapes.

That something else is odd about this assumption of wholeness or unity insofar as it informs landscape studies was suggested by Jackson when he wrote, "The well-designed city is one where everyone feels at home."[17] This kind of normative view of cities downplays and draws attention away from significant landscape disjunctions, from the immense human and material diversity of a landscape that cannot be known merely through reading and armchair speculation. Indeed, Jackson's design renders diversity pricey, almost valueless. It is not surprising, then, that it is hard to find gritty and evocative down-home landscape descriptions in the writings among senior cultural geographers—or any of their churchly followers, for that matter.

What is, with few exceptions, so striking about Jackson and others I have mentioned is how distant, cold, abstract, and impoverished their landscape descriptions seem, how bereft, for example, they are of the voices and ethnic richness of cities and towns they claim to know.[18] Lacking is a robust sense of enthralling localisms, idiosyncrasies, peculiarities, diversity—all those finely woven landscape descriptions that make one feel like he is there—that one repeatedly encounters in the writings of people such as Edward Abbey, William Least Heat Moon, Barry Lopez, or John McPhee.[19] That there is not all that much attention to descriptions of the built environment in these writings of people, that these writers seldom follow established canons of scholarship (neither do Meinig and Jackson consistently, notwithstanding claims, implicit and otherwise, to the contrary), that they are not formally inscribed in the geographic profession: none is reason enough to ignore the resonant message their writings convey. What is striking about these writers is not just their felicitous language, the way their sense of dialogue and pace animates landscapes, but also their love of one-on-one engagements with places and people, their firm denial that the world beyond their minds fits together like organs in an animal, and their fascination with difference and diversity—with individuals.

For J. B. Jackson, the assumption of landscape unity appears to be an ad hoc justification for a kind of semireligious nineteenth-century teleology and an often-intemperate ad hoc functionalism. Jackson's vision, for example, found expression in a 1968 piece in *Landscape* titled "Life-Worship." There he wrote, "We can foresee in the not very distant future an America transformed. Our cities will be no more splendid than they are now, but they will be clean and cheerful and safe." Elsewhere Jackson wrote, "The separation [of man and nature] . . . is not primarily a physical one, an inevitable outcome of modern urban existence; it is a separation incorporated in our dichotomous way of thinking. It is a nineteenth-century aberration and in time it will pass."[20] These words are from an idealist, someone detached from and apparently ignorant of people in the ordinary landscapes of towns and cities as we know them—certainly as I know them.

A historical analysis of anything is valuable, and more often than not indispensable. But history as a written or oral record is often sketchy, incomplete, and full of cursory and half-baked observations and opinions. One hell of a lot of history offered as reliable guides to reality in the present would be scoffed at, judged no better than pedestrian journalism not worthy of the serious historian. Furthermore, history can be used in ways that suggest excessive authority, particularly if a cited author holds an esteemed place in the history of letters. In the hands of a deft scholar or writer, history can be used to purportedly explain a pattern—present or past—when in fact it may do no such thing. It does not explain not only because the questions presently of interest were not asked as such when a past present was unfolding, but also because there may be a fundamental disjunction between the point at which the history of this or that landscape leaves off and fundamental change occurs, a change so dramatic and rending that history is at best prelude and context, a rather meager guide to meanings of the moment. But equally dangerous is the temptation to seek in the lone articulate individual in history a full-blown explanation, a universal, or perhaps a reversal of a claim about place and space that may or may not be true. Thus, for example, David Sopher, in "The Landscape of Home," disputed Yi-Fu Tuan's assertion that Americans "lack attachment to place."[21] Sopher turned not to a survey or to long, personal foot-stepping experience in American landscapes to take issue with Tuan—who himself has virtually never engaged the people about whom he has so often glibly generalized—

but to a black writer from Harlem who remarked that when he got off the Duke Ellington "A" Train, he felt at home.[22]

Attachment to place is not a minor concern to students of landscape, and the issue of who is and who is not attached to this or that place at this or that point in history cannot be resolved—and can be only muddied—by respected scholars citing a couple of celebrated writers who have caught their eyes, writers perhaps no more intimate with locals and local landscapes than Sopher or Tuan. None of the above is to gainsay the value of history. Rather, the point here is that history can too easily become a retreat, an apparently unassailable safe house for staying out of the present, relegating local views to a secondary or nonexistent epistemological category. Relying too much on history when one is trying to explain the present can make a landscape seem saturated with symbolisms and discourses and much more rooted in the past than can possibly be justified after soaking oneself in local views of a present landscape.

If one buys the idea that locals and their views are, at best, of secondary interest, and that intimate familiarity with the historical record and closely observed material artifacts are sufficient conditions for understanding ordinary landscapes, then it is not surprising that one might well be more interested in similarities than in differences, in glossing over what distinguishes one landscape from another. That one incorporates this assumption into his epistemology may be less a matter of constitutional aversion to place-to-place differences—to what makes geography geography and not sociology—than it is a logical outcome of aversion to living and breathing ordinary landscapes.

A captivating essay written by Jackson very early in his tenure as editor of *Landscape,* in 1952, starkly revealed attachment to this proposition.[23] In what Jackson titled "The Almost Perfect Town," he created an imaginary place called Optimo City. He convincingly detailed the town's history, its gridiron layout, and the striking differences between an imagined South Main Street and North Main Street. It all seemed perfectly real, and you wanted him to go on and on to get you further inside the soul of this amazing place you had never been to.

> South Main Street, which leads from the square down to the river, was too steep in the old days for heavily laden wagons to climb in wet weather, so at the foot of it on the flats near the river, those merchants who dealt in farm produce and farm equipment built their stores and warehouses. The

blacksmith and welder, the hay and grain supply, and finally the auction ring and the farmers' market found South Main the best location in town for their purpose—which purpose being primarily dealing with out-of-town farmers and ranchers.[24]

A little further on he wrote, "North Main, up on the heights beyond the Courthouse Square and past the two or three blocks of retail stores, is (on the other hand) the very finest part of Optimo. The northwestern section of town, with its tree-shaded streets, its view over the river and the prairie, its summer breezes, has always been identified with wealth and fashion as Optimo understands them." Jackson had more to say about his imaginary town, taking note, for example, of its "Latino shacks under the cottonwoods and next to the river," its "tin roof porches," its "dignified business section," the air that on Saturday night is "full of pigeons and floating candy wrappers, the flat strong accent erroneously called Texan."[25]

What Jackson invented is fascinating, engrossing, pleasing, fetching, in fact so much so that when, many years ago, I finished the piece, I looked through several atlases, hoping that Optimo City was a real place and that I could get in my pickup and find out how the place had changed in the many years since Jackson wrote about it. Jackson had convinced me that despite all my travels in America and the Southwest, this place that I'd not yet enjoyed was real. But, alas, Jackson left absolutely no doubt about his purpose, namely, that he did not fundamentally care about differences, diversity, the abiding richness and singularity of unique places. He opened the essay by stating, "Optimo City is not one town, it is a hundred or more towns, all very much alike, scattered across the United States from the Alleghenies to the Pacific, most numerous west of the Mississippi and south of the Platte." And then he ended the piece by saying, "There is another Optimo City fifty miles farther on. The country is covered with them. Indeed they are so numerous that it sometimes seems as if Optimo and rural America were one and indivisible."[26] I do not know whether Jackson has traveled more than I have on the major and minor highways of America, but I have traveled a good deal by car and other means in this country and cannot identify with Jackson's confident claim that America is covered with Optimos.

The position that similarities are more important than differences led Jackson to some odd conclusions. For example, he said:

I have come to the point where instead of trying to establish distinctions between landscapes, I try to discover similarities; that is one of the differences between the professional and amateur traveler: the professional searches for (and finds) differences, and is partial to what might be called a kind of academic romanticism: the establishing of distinct categories. The amateur, on the other hand, is more concerned with finding similarities, with perceiving the universal which presumably lies behind diversity.[27]

What is puzzling here is that if anyone could be tagged with academic romanticism, then it would be Jackson, especially when clothed in teleological underclothes, functionalist pleading, and hard-to-find harmonies between humans and their environment. It is *not*—contra Jackson—"professional travelers" who are likely to wax poetic about nasty humans and their defilement of their environment, but rather detached academics who describe and categorize landscapes from afar: from what they learn in books and from what they see from their automobiles or train or plane seats, or atop their Harleys. But more serious was Jackson's unexamined assumption that "universals," or broadly based generalizations about American landscapes, were more than figments of the imagination, and they could be derived by ignoring all the diversity that Jackson and others who knelt at his altar ignored. The liquid history that Jackson wrote, his ability to gather disciples and form a school, his admirable tenure as editor of *Landscape*, his apparent intelligence, his professed interest in landscapes both European and American: none was reason enough to suppose that he could, with any regularity, make universal statements about landscapes that are both engaging and right. Nor could such be expected from the best and the brightest in geography or any other academic discipline. Meinig was quite right in saying of Jackson: "All is assertion and argument, nothing is documented or formally demonstrated; much is observed, nothing is measured."[28] What Meinig should have added is that this portrayal was, despite the occasional paraphernalia of scholarship, also an apt description of himself as a student of ordinary landscapes.

A good example of facile overgeneralization common to Jackson admirers could be found in a 1980 *Landscape* piece titled "Rootedness Versus Sense of Place," by Yi-Fu Tuan. Tuan asserted, "In contemporary American culture there is a strong longing for roots, for recapturing or restoring a sense of place."[29] This claim is worthy of at least a couple of Ph.D. dissertations, to find out if it is indeed true, to disaggregate it, to measure the strength of the longing. Tuan, unfortunately, offered no data whatsoever—

nothing more than spurious passages from a poem and irrelevant examples of rootedness among Congo Pygmies, the Tasaday, and the Kung bushmen. Indeed, one could find more reliable and relevant data on the front page of *USA Today* were the issue addressed.

Meinig's strategy to laying claims to universality about landscapes has usually taken a different turn, one rather foreign to Jackson's nonacademic, "amateur" proclivity. Meinig again and again has cleverly protected himself by retreating into that secure academic sanctuary called "ideal forms," "model landscapes," or "idealized versions of an actual landscape."[30] Just when you want to take issue with his characterization of, say, New England landscapes or the "Main Street of Middle America," Meinig gives you the poolroom hustle and a fast slip: "This is of course a projection from an actual landscape and society."[31] But what, Don, does the landscape of New England or the Midwest really look like, feel like? How much does your idealization have to be disassembled and reordered to be able to confidently identify the real thing, one general enough to ring true, to not do disservice to locals and the local worldview? Like bad science that gets published in reputable journals, seemingly benign models and idealizations are baneful. They are baneful not just because they are a form of academic gamesmanship, not even because one need never worry about being nailed for matters of fact and on-site reason, but because they get entrenched in the written record and become a solid kind of truth. Such a "truth" is terribly difficult to erase or bury. Worse, fascination with such "truths" forestalls the day when the local view will be heard.

Appearances, exhortations, and scholarly treatises to the contrary, Jackson (now well into his everlasting-sainthood tenure) and his principal apostles have not been the best exemplars of the practice of a hardy, down-to-earth human geography or of the study of ordinary landscapes. They have long been tracking, unchallenged, in the wrong direction. What is needed is a philosophical revamping, a reordering of priorities. It is, I think (obvious by now), primarily about the mouth and getting out there, about shedding the academic pretense and the preening and about listening to voices other than your own. It is about using the eyes but not being arrogant about what they can do.

Part Four ◎ Censured

The most thorough analysis of moral obligations is unquestionably that of Panaetius, and on the whole, with certain modifications, I have followed him. The questions relating to this topic which arouse most discussion and inquiry are classified by Panaetius under three headings:

1. Is a thing morally right or wrong?
2. Is it advantageous or disadvantageous?
3. If apparent right and apparent advantage clash, what is to be the basis for our choice between them?

—Cicero, *Selected Works*

Assistant professor

A Brief Meditation on Censorship

Confront improper conduct, not by retaliation, but by example.

—J. Foster

THE censorship of ideas in a democratic society is abhorrent. It is particularly abhorrent in universities in democratic societies, universities in which the principle of academic freedom is held sacred. The principle of academic freedom, of course, is but a specific expression of our nation's most revered amendment to the United States Constitution. It is censorship within the American university, and, in particular, within the discipline of geography, to which I direct my thoughts. But before getting specific—coming forth with the details of some cases involving myself—it is worth briefly exploring a few categories and relationships that litter the censorship terrain.

To be significant, worthy of more than passing attention, a censor must purport to represent or speak for a group, an institution, a body of some sort that speaks either for or to others, and in numbers that are not trivial. Nontrivial does not mean large numbers. It could refer to a small class of students, or a university department. But then it might just as well apply to a professional academic body of several thousand members, a university of tens of thousands. In each instance, someone is attempting to disseminate an idea to one of these institutionalized aggregations, and that person or persons—who may be representing aggregations of people that are formally or informally institutionalized at various scales—is either prevented from doing so or, in some significant degree, at least temporarily thwarted from doing so.

Whether the attempt to censor an idea is successful in whole or part is somewhat independent of the strength of the desire of one or more people representing the aggregate to do so. The reasons here are many: the force or significance of the idea, the asymmetrical power relationship between the

. disseminator and the censoring party, the ability of disseminator or censor to marshal forces either for or against the idea invading the body politic, changing attitudes about the idea, and changes in the mind-set or persistence of either the disseminator or the censoring party. Other reasons might be imagined.

It is not obvious who in an organization will act as the initial censor, or who will be intimately involved in the process of censorship as an attempt to disseminate an idea is made. To be sure, there are kinds of people in all organizations who are the first to be confronted by a new, different, or unacceptable idea that may get into the body politic. These people may be lodged somewhere within the managerial structure of the organization, but then again they may be numbered among the rank and file. In the latter instance, it may simply be fortuitous that someone in the rank and file was, so to speak, at the wrong place at the wrong time, and thereupon confronted with an idea that that person took to be unacceptable to the group to which that person belongs. Here, among the rank and file, the power of censorship—censorship of a nontrivial nature—cannot be exercised in any significant way. But the idea as threat can quickly be passed on to someone within the managerial ranks, someone who has or purports to have the power of censorship on behalf of the organization.

From these few observations, it can be seen that the source of trouble does not necessarily reside, *in the first instance*, within the managerial or ruling ranks of an organization. Thus, a distinction needs to be made between a first instance of censorship, one that may be relatively innocuous, and confirmation of this first-level censorship, one—to repeat—that may have little force or effect per se. Confirmation within the managerial or ruling ranks of an organization, of course, takes censorship to a new and more profound level.

In this complex arena, there is no clear and predictable relationship to be imagined a priori between the purveyor of an idea and the censor, censor here meant to include both the organizational representative and the organization being represented. Although the person attempting to disseminate the idea may in one case be intimately linked with the idea such that the two cannot be easily separated, in other cases the person as such, insofar as the censor and the organization behind the censor are concerned, may be, pure and simple, no more than a messenger. Out of this brief sketch of possibilities, then, one might imagine three types of situations. In one, only the idea is at issue or is being censored. In another, the idea is not really at issue; rather, it is the person who is being censored—the idea is merely pretext. And in the third case, it is both idea and person that are being censored; there is something unwelcome or distasteful about both of them. These distinctions have some further implications when thinking about or examining a particular case of censorship.

Preston James, twice (once honorary) president of the Association of American Geographers and the self-styled maker of geography at Syracuse University, often remarked during my student days at Syracuse that someone ought to do a geography of prostitution. It was, he opined, a fine topic to tackle, especially in Latin America. After all, James knew all about it from experience, and because—by his own reckoning—he knew more about Latin America than any living person.

Preston James was not on my mind when I decided to go to Nevada in the summer of 1973 to see what I could find out about Nevada prostitution. Rather, I'd brought to mind a long conversation I'd had with a madam in Battle Mountain, Nevada, in the summer of 1967. In that run-down brothel in Battle Mountain, then known as the Green Lantern (subsequently flattened for a more upscale version under a different name), I was into my second or third beer at the bar when the hefty and garrulous madam who called herself Shirley began giving me the lowdown on laws and regulations regarding prostitution in Battle Mountain, Elko, and Winnemucca. She talked authoritatively about how they applied to her and her "working girls" and their pimps and boyfriends. With less authority, she went on about how laws and local regulations applied elsewhere in the state.

The research paper that resulted from this 1973 fieldwork, "Prostitution in Nevada," was published in one of geography's premier journals, the *Annals of the Association of American Geographers*. It was published less than a year after it was written, and though the journal has always touted that all its articles are refereed, "Prostitution in Nevada" was judged by no one other than the editor. The reasons, I have long thought, were twofold. The editor from 1970 to 1975, John Fraser Hart, a professor of geography at the University of Minnesota and some years later, in 1979, president of the Association of American Geographers, claimed that it would be hard to find two or three geographers to recommend publication. No one up to this point in the history of American geography had ever published an article in a major journal on a sex-related topic, and academic geographers have long had a reputation of being among the most conservative of social scientists. John Fraser Hart was, by numerous measures, conservative, yet he had a strong iconoclastic streak. Publishing "Prostitution in Nevada" was an act of defiance, an "in-your-face" gesture that he felt would create the kind of gossipy buzz and oh-nos that he so relished.

There was plenty of grumbling and outrage. One of the most prominent and well-known geographers at UCLA, and editor of the *Annals* from 1964 to 1969, Joe Spencer, had a veritable fit over the article's publication. He, like others, felt that this research was precisely the kind that would further tarnish geography's already fragile image.

Not long after publication, Fraser Hart wrote to me and said that I could expect a lawsuit in the near future. He claimed that a professor of geography at a major West Coast university had brought home the issue with my article in it, left it on the dining room table, and his wife had picked up the issue and thumbed through my article. And there, lo and behold, she saw a Cadillac parked in front of one of Winnemucca's whorehouses, the Players Club. Alas, it was her Cadillac! She knew it beyond doubt, because the license plate number could be read. Fraser Hart said that not only would both of us be sued for defamation of character, but we were also directly responsible for the breakup of a marriage and would pay accordingly.

Before long, Fraser Hart wrote to me to say that the whole story was a fabrication. Ha, ha! Then he added that both of us should really have been more careful in examining the photographs before including them in the article. He was right.

By my estimate, about three of four practicing academics (not just geographers) who read or knew of this article, and may or may not have known me previously, would upon meeting me ask (often chuckling while doing so) exactly *how* I did the research. How many whores had I slept with? Were they any good? Rare was the academic geographer (or academic of any stripe, for that matter) who had seriously read the article and had even a semiserious question for me about something that I had written about.

Time would show that this single article greatly skewed almost every judgment of me and my character and how people approached or reacted to me. With this single article, I had walked a long way toward being blackballed in academic geography. Much of what was said to me in person was meant to convey the message that, for an academic geographer, I had ventured where others only dreamed of going when drunk, or beyond the gaze of a public eye. Prostitutes and brothels were trouble, and getting close to whores was, I was to understand, though titillating and exciting, also forbidden. With it came a price, an indelible mark on one's image. As a friend on the geography faculty at the University of Wisconsin at Madison would say to me several years after the article appeared: "You have dared to associate with outcasts and criminals—sexual ones no less—and now many in the profession associate you with the kinds of behavior for which prostitutes are infamous."

Colleagues in the geography department at the University of Texas at Austin where I was an assistant professor of geography and Latin American studies at the time the article appeared were strongly critical of my right to have researched and written it. They were critical not because they thought the research was shoddy—I never heard this complaint—but because of the subject matter. I had written on a taboo topic, and I should have known better. As one elder statesman in the department by the name of George Hoffman censoriously said to me as he wrapped an arm around my shoulder in the hall outside my office one day, "You will never do this kind of research on those awful women again, will you?" Over the years, I would find that he spoke for a great many academic geographers, many of whom advertised themselves as liberals—yes, liberals—so many of whom are now numbered among the appropriately labeled "liberal fascists."

From time to time, I would discover that the attitude toward this kind of research was simply chilling, and quite costly. When, in 1979, I was to get an academic position at the University of Illinois at Champaign-Urbana, I learned that several members of that geography department were opposed to my appointment because of my research on prostitution. None of my other published research—and there was plenty of it—was an issue, and quite the contrary. At this time, I had nearly finished a book-length manuscript that would become *The Immoral Landscape: Female Prostitution in Western Societies,* and when it became known by the academics at Illinois, it only solidified their certainty that I was unfit to be a member of that department. The chair of geography at that time, Arthur Getis, proclaimed that he could not imagine why I would want to do such research because, he told me to my face, there was "nothing more to be said about prostitution that wasn't already known." How Getis could possibly have known such a thing because anything having to do with social, cultural, or political geography was leagues from his professional interests (largely, sterile mathematical modeling), I do not know. More telling was a comment received directly from one of the department's flag-waving liberals. He informed me that the department could not possibly afford to hire me, because should the state legislature in Springfield hear of my research, they might well close the geography department at the University of Illinois. Other professors in that department were too cowardly to even tell me what they thought; I heard about their censorious feelings only through gossip and innuendo. There may have been other reasons I did not get the job I was promised at Illinois, but whatever those other reasons, I was certain then, as I am now, that my research on prostitution had indelibly stamped me as undesirable, and made me far too big a risk to hire.

By the end of the twentieth century, research on sex in geography was commonplace—sort of. It was now perfectly okay to write from every imaginable angle about lesbians and gays and their sexual behavior.[1] But it was still quite another matter for a male who would dare to do research on prostitution. After all, there could not possibly be any motive for doing so other than to have sex with the women, and this kind of behavior (imputed and imagined, never documented) is, as those moralizing police on the fascist academic left are quick to remind us, beyond the pale of what is acceptable.

In Revolutionary Mode

> What is the nature of *Antipode*'s Marxism? The answer is broad,
> eclectic, and tolerant. And of the politics we espouse? Most Marxists
> in geography believe in the most democratic of socialisms, the
> extension of democracy into all institutions in society. . . .
> Geographers should be reminded that it is this belief which they
> oppose when they fire us, refuse us tenure, and otherwise express
> their low opinion of us.
>
> —Dick Peet, self-styled Marxist-anarchist

THE occasion was the 1977 Annual Meeting of the Association of American
Geographers, and I had sought Dick Peet to inquire when he was going to
publish a paper written by me and my wife, Nancy Burley. The paper, "Geography and Natural Selection—Revisited," attempted to explain the neo-
Darwinian perspective on evolution, as well as several implications of
sociobiological reasoning for geography.[1] Some months before the annual
meeting, Peet, as editor of *Antipode: A Radical Journal of Geography*, had
had the article reviewed. The reviews were favorable, and Nancy and I had
gone along with the suggested changes and sent the paper back to Peet. We
believed that the paper's appearance in *Antipode* was a foregone conclusion.

When I saw Dick late that morning seated at a long table among Marxist friends peddling *Antipode* memberships and back issues and other radical material, I asked him when he expected to publish the paper. He
responded by saying that he didn't know, but that it was important for me
to speak to his "comrade" Jim Blaut as soon as possible. Fine, I said. I
waited for Blaut to come around.

Presently, Jim Blaut appeared. Before I had a chance to hitch my pants
and jump off the table I was sitting on, Blaut went into a spitting tantrum
that sociobiology serves only the "establishment and the status quo and

171

the oppressors," and anyone who embraces sociobiology—in this case meaning anyone who wrote about it—is "counterrevolutionary." As I was to understand, doing so made me the '70s version of the '60s short-haired pig. Blaut then went on to say that *Antipode* was no place for my paper, and he sure as hell hoped that Peet had the good sense not to publish it. It's nonsense, not good for anything, he said.

That may be, I said. But it's been refereed and accepted, so you'll have a chance to air your opinions in print.

His face grew large and ugly, and he muttered that I was talking nonsense. He blathered on, and then Dick Peet more or less came to my rescue. Dick said, Yes, I've accepted the paper. Jim Blaut screwed up his face, as if I'd taken a sledgehammer to a hometown statue of his hero Lenin.

Have you read the paper? I asked Blaut.

No, and that's irrelevant, he shouted, loud enough to make a passerby wonder who had just been raped.

I insisted that it would be in his interest to read the paper before he shot off his mouth. After all, I reminded him, as a university professor it was the very least of his responsibilities. I tried to get across the message that little was more insulting than to be told that you'd written a piece of twaddle by someone who hadn't even read a line of the paper. (However, I was not unfamiliar with this macho academic tactic. One of my dissertation committee members had pulled the same trick on me when my Ph.D. dissertation was at issue.) But Blaut was interested only in hearing himself decry sociobiology, in telling me what a fool I was for going anywhere near it, in telling me more about my politics than even my mysterious id knows. Finally, in a rare moment when he stopped blabbering long enough for me to get in a few words, I asked him if he had read any of the sociobiological literature.

I can't be bothered, he said.

So you've not read *any* of it? I insisted.

He repeated what he had just said, and he reiterated that he had no intention of doing so and asked what was wrong with me that I didn't see that he and his comrades were in a ghetto just like Puerto Ricans, and when you're in a ghetto fighting the oppressors, you can't afford to listen to wrongheaded, reactionary claptrap, especially by a dimwit who clearly doesn't know what he's talking about.

And so as I patiently listened to Professor James Blaut, now and again trying to squeeze in a question or make a point through his showering storm of rhetorical bile and heavy spit, I wondered how it was possible that this hulking, self-proclaimed savior of the Puerto Rican lumpen proletariat and 141 subscribers of *Antipode* had ever been certified for anything more than working as Big Man on a commercial garbage truck. Through all of Blaut's heavyweight shouting, Dick Peet stood quietly on the sidelines, occasionally moving around us like an amateur referee. He said little. What I remember most is the look of fear and trepidation on Peet's thickly bearded face, the way he lowered his balding head as if taking the sacraments on the steps of St. Peter's. That day, Dick Peet struck me as a pitiable prisoner to ideas and enchanting words such as *contradiction, dialectic,* and *revolutionary* that he didn't really understand. I also remember thinking that his concept of the real meaning of revolution was probably so weak that he wouldn't know what to do if on a late-night stroll he bumped into a bum buzzing with the D.T.'s. I wondered: Would Peet or Blaut or any of the discipline's revolutionary hotheads eager to proclaim their allegiance to Marx, Lenin, and Stalin give a down-and-outer a room for the night in their bourgeois homes, just down the hall from their cuddly and well-protected kids? Would any of them give him a meal, a shower, a clean shirt, and, in the case of rad-left academic professorial sorts such as Peet and Blaut, more than pocket change from their considerable academic salaries? I seriously doubted it.

It was just about a month later when I received a letter from Peet saying that "after listening to certain objections made to the paper by such comrades as Jim Blaut, I'm still committed to publishing it, but now I want to have a couple of replies published at the same time." He didn't say who he'd get to write the replies (replies being something of a first in *Antipode* with our paper), but I figured it was safe to assume that they'd be comrades. To Dick's credit, he did say that I'd have a chance to write a rebuttal.

Okay . . . mere change of rules, I thought. Nothing to get excited about.

In the same letter, Peet went on to say that he couldn't get the replies and my rebuttal together in time for the next general issue of *Antipode.* "However, you needn't worry," he wrote. "The paper will appear at the end of the year or early next year [1978]." He ended by asking me to read the first three chapters of P. A. Kropotkin's *Ethics.* "This more or less states my own point of view on the matter."

In a collection of papers edited by Peet that appeared that very year, *Radical Geography: Alternative Viewpoints on Contemporary Social Is-*

sues, Peet dedicated the book to Kropotkin.[2] He included a quote from *Ethics.* Part of that quote is worth noting.

> Thus science and philosophy have given us both the material strength and the freedom of thought which are required for calling into life the constructive forces that may lead mankind to a new era of progress. There is, however, one branch of knowledge which lags behind. It is ethics, the teaching of the fundamental principles of morality. A new, realistic moral science is the need of the day—a science as free from superstition, religious dogmatism, and metaphysical mythology as modern cosmogony and philosophy already are, and permeated at the same time with those higher feelings and brighter hopes which are inspired by the modern knowledge of man and his history—this is what humanity is persistently demanding.[3]

I concluded that Peet wanted me to get a couple of messages: First, Marxism and radical thought are moral, the kind of "constructive forces that may lead mankind to a new era of progress." Sociobiology is immoral, a pernicious force that is retrogressive. (It's worth noting that at this point, Dick Peet had no more inkling of what sociobiology was about than did Jim Blaut.) Second, Marxism and radical thought deal with "modern knowledge of man and his history." Sociobiology, he probably would have contended, does not. (This belief is rather odd, as most Marxists do a rather shoddy job of keeping to historical facts, and their understanding of the evolutionary vessel of history is, with few exceptions, appalling.)

But the real rub was how Dick Peet was ever to reconcile his wholehearted embrace of Kropotkin, and casual dismissal of sociobiology, with his fervent religiosity—his love of "superstition, religious dogmatism, and metaphysical mythology." But then it is I, perhaps, who should have tried to understand Peet's predicament, for surely the religious dogmatist cannot stand outside himself and possibly understand what it means to be either religious or dogmatic.

Before long, I got another letter from Peet, this one informing me that one of his comrades, Professor Phil O'Keefe, was taking over *Antipode* while he went off to Australia to ruminate on how to resolve certain unresolved Marxist issues. "The journal is in good hands," Peet wrote, "and Phil will publish your paper in due course."

I wrote to O'Keefe several times after he took over the reins of *Antipode.* I repeatedly asked when the paper would be published.

"The time is not right," he wrote back, once. He didn't respond to my other letters. The paper was never published by *Antipode.* Nor has Dick Peet ever been willing to address the reasons for it, or why he has behaved the way he has. Rather, when it comes to criticism of his behavior and that of like-minded comrades, he responds by letting it be known that I am "crazy" and ought to be ignored—and no doubt sent to a gulag, were he to have other than paper power among comrades.

An Anatomy of Academic Censorship

A professional organization devoted to furthering truth.

—Mission statement of the Association of American
Geographers

BOOK censorship has a long and infamous history in America, and even today there are K–12 schools around the country that will not permit such books as *Catcher in the Rye* and *Slaughterhouse Five* in their libraries.[1] But American universities are different—or so one would think. They are allegedly populated with liberal professors, they profess to seek truth and insight into the human condition—wherever such seeking may lead—and in the academy, no principle is more jealously guarded than that of academic freedom, the right to openly express one's views and to explore subject matters that are considered not only unpopular but even subversive. One might well argue, and with some justification, that recent concerns with "political correctness" are undermining the sanctity of academic freedom.[2] Be this as it may, the principle nonetheless remains absolutely central to the very definition and mission of American universities. Furthermore, academic freedom is but a particular expression of one of the most cherished doctrines of a democratic society: the First Amendment of the United States Constitution. Justice Brennan came right to the point when he noted:

> [Academic freedom] is . . . a special concern of the First Amendment, which does not tolerate laws that cast a pall of orthodoxy over the classroom. The vigilant protection of constitutional freedom is nowhere more vital than in the community of American schools. The Nation's future depends upon leaders trained through wide exposure to that robust exchange

of ideas which discovers truth out of a multitude of tongues, [rather] than through any kind of authoritative selection.[3]

It is a pivotal and frightening moment, then, when an academic discipline engages in censorship, the willful suppression of information that will allow its professorial membership to buy or not buy a book on the grounds that the subject matter of the book "might offend some members of the profession." That it happened at all in an academic discipline as recent as 1982 seems surprising indeed. That it occurred without anyone involved in the censorship having seen the book is seemingly incredulous, given the importance attached to careful scholarship, the careful weighing of evidence, and "objectivity" in the academy.

In 1975, I began serious research on prostitution, for a book that would eventually become *The Immoral Landscape: Female Prostitution in Western Societies.* I visited several major libraries for sources, including the world-renowned library in the Kinsey Sex Institute at Indiana University and the newspaper and manuscript collection maintained by Coyote, a prostitute union in San Francisco.[4] For the book's lengthy appendix on the evolutionary basis for prostitution (written jointly by me and my wife), I examined every society (more than three hundred) then included in the Human Relations Area File. Research for the body of the book included fieldwork in Montreal, San Francisco, Oakland, New York City, and every city and town in Nevada where brothels were found in the mid-1970s. The book has more than eight hundred bibliographic references, some eleven hundred footnotes, and more than eighty-five maps and tables, most of which were original. Prior to acceptance by the publisher, Butterworths, then and now one of the more reputable and long-lived publishers in the Western world, *The Immoral Landscape* was reviewed by three people with academic credentials in sociology or sex research or both. The cost of the book to me and to the publisher from commencement of research until its appearance in 1981 was approximately thirty-five thousand dollars.[5]

Like most publishers, Butterworths asked me as author to identify major markets for the book. Because the book is inherently geographical, and because my principal reference group at that time was academic geographers, I suggested that Butterworths buy the mailing list of the Association of American Geographers and send fliers to this target group.

The AAG is no fly-by-night organization. Since 1910, and to this day, the AAG has been the major professional organization for geographers, ones who are academic and a great many who are not but work for the government and in private businesses around the world. At the time of the censorship, the organization had some six thousand members.

On the morning of June 4, 1982, I called Butterworths in Toronto (where the book had been edited and printed) to inquire about the promotion of my book. Laura Howard, Butterworths's promotions assistant, informed me that they were having unprecedented problems. She had received a letter dated March 15, 1982, from Patricia McWethy, executive director of the AAG. Howard was told by McWethy that the AAG would not sell its mailing list for purposes of sending out a flier announcing the availability of *The Immoral Landscape.* In her letter to Butterworths, McWethy gave only one reason for the decision: "I hope you understand that a number of our members would be offended by our participation in the promotion of this publication."[6]

Within minutes of talking with Laura Howard, I called McWethy and confirmed with her that she had indeed refused to sell the AAG mailing list to Butterworths. I also confirmed that she had done so for the sole reason given in her letter to Laura Howard. I asked McWethy to change her mind. I explained that her action was a blatant violation of the First Amendment of the United States Constitution and the principle of academic freedom. Cognizant that McWethy did not have a research degree (she had a master's degree in business), I told her that the principle of academic freedom was the most sacred tenet in universities. I asked her if she had seen my book. She said she had not. I asked her if there was anything offensive about the flier that she had been sent by Butterworths. She said there was nothing offensive about it.

I persisted in my demand that McWethy sell the mailing list to Butterworths, arguing that she had made a grave mistake and that I would do everything in my power to get her decision reversed were she not to do so. She said there was no possibility that she would change her mind. She said that she disapproved of research on prostitutes, and she knew "of some people" in the discipline who had objected to my article on prostitution in Nevada that had been published in the *Annals of the Association of American Geographers* in 1974.[7] She did not say why they had objected, and from the conversation, I gathered that she did not know. I asked her if she thought the article was unscholarly. She said she did not know; she had not read it. Nowhere in our discussion did she say a word about AAG mailing-list policy or whether I was currently a member of the AAG (which at that time I was not). At this point, I demanded that McWethy call Butterworths and tell the appropriate people that she had changed her mind. She hung up on me.

Pat McWethy had previously made it known that the AAG mailing list was routinely available to reputable academic publishing houses such as Butterworths. On March 18, 1981, in answer to a letter requesting information from Robert Lane, Butterworths's sales manager, she wrote that approving a book for which a mailing list would be sold was a "routine matter" done by her. At that time, she gave Lane the cost of purchasing pressure-sensitive mailing labels, and she noted that there were fifty-five hundred names on the list. She also said: "It is not possible for us to enclose a brochure with one of our journals, however, you may wish to advertise in the *Annals* and/or *The Professional Geographer.*"

Was there anything in the flier that could be construed as offensive, that would have given McWethy, or anyone else, reason for refusing to sell the mailing list to Butterworths? All the flier contained was a chapter outline, a brief description of what the book attempted to cover, a paragraph from the first chapter in which I gave reasons that prostitution is a social problem with no obvious solution, the price of the book, and where it could be purchased. The flier took note of the scholarly nature of the book and the ground that I covered:

> *The Immoral Landscape* is a study of prostitution that draws on sociology, anthropology, geography, biology, and law. Using a wide range of historical and contemporary materials, including data collected in the field, Richard Symanski examines the history of exploitation of prostitutes. He shows the many ways in which the state, prostitutes, and others have used geographical strategies to solve their problems. Numerous original maps, graphs, and photographs are used to illustrate the author's points.

In the afternoon of June 4, I wrote a long and strong letter to the National Council of the Association of American Geographers, the executive council ultimately responsible for McWethy's actions. The council had the power to reprimand or dismiss McWethy. I explained McWethy's position on the book, and I highlighted the fact that it was an abridgment of both the First Amendment and the cherished principle of academic freedom. I said that I was outraged by her action, and that because she had insisted that she would not change her mind, she ought to be asked to resign. I asked that the council reverse her decision immediately.

I sent copies of the letter to approximately seventy people, all but a few of whom were geographers with tenured university positions. Copies were sent to eleven past, present, or incoming presidents of the AAG; seven past or present editors of the *Annals,* the *Professional Geographer,* the *Geographical Review,* and *Economic Geography;* and more than a dozen department heads. The letter was sent to Richard Morrill, professor of geography at the University of Washington and president of the AAG; John Adams, professor of geography at the University of Minnesota and vice president and incoming president of the AAG; and all of the people then serving on the AAG's national council.

I sent as many letters as I did because I feared that were only a few people to hear of the censorship, it might be months before the council would take action, a delay that would surely slow the book's promotion, and might even kill any possibility of decent sales. I also did not want the censorship situation misrepresented by word of mouth.

Six days after my letter of protest was mailed, John Adams wrote a letter to me, with copies to the AAG council, McWethy, and Butterworths. He informed me that he had talked with "several council members to verify the long standing policy of the executive office regarding requests of this kind. The presumption has been that the list should not be sold unless there are compelling reasons from a 'services to the membership' standpoint, for doing so." Adams went on: "There are IRS regulations that discourage such sales as revenue raising measures, unless, again, the activity is an important service to the membership."

Professor John Adams had more in mind by way of assuring me that the action taken was legitimate. "Many members," he wrote, "object strenuously to the use of the list for advertising purposes. They have the right to object and expect that their voices will be heard. Publishers, airlines, travel bureaus, rental car companies, and insurance companies have all tried to buy our lists and we usually turn them down. They have no right to the list. It is our decision whether to sell it." Adams ended his letter to me with the following line: "Unless the council changes the policy—which I do not think they are inclined to do—the Executive Director will continue to make case by case determinations of whether or not to sell the list. Until that policy changes, I support the present policy."

By June 14, I had received almost two dozen letters from academic geographers protesting McWethy's decision. Many of these letters were addressed to McWethy, and to Morrill and Adams, with copies to me. Several geographers called or wrote to me and told me that they had made telephone calls to McWethy, Morrill, Adams, or others on the national council inquiring about my allegations. An ample sample of the letters follows.

From Ross MacKinnon, chairman of the geography department at SUNY-Buffalo, to Pat McWethy:

> If the events Symanski relates in his letter are accurate, you can count me among those who are offended by your decision. I have not seen the book, but the flier leads me to believe that the topic, distasteful as it may be to you, is being dealt with in a scholarly, responsible way. Poverty and racial discrimination are equally distasteful, at least to me, but they are also valid topics for research by geographers.

From Robert McNee, one-time national councilor, former chairman of the geography department at the University of Cincinnati, and former director of the American Geographical Society, to Pat McWethy:

> I have not read the book (though I have ordered it). However, I have read the flier and it is in the best of taste. Or don't you agree? If not, please explain. If [Symanski's] charges are true, the public relations damage to the AAG could be very great. I have a framed copy of the First Amendment to the U.S. Constitution in my office at the University of Cincinnati and another in my living room at home. I know that a great many others in the AAG are just as passionately devoted to freedom of information as I am. Surely, as a scientific professional, you share our belief in freedom. As a longtime member of the AAG and a former Council member, I am much concerned that the AAG not appear anywhere at anytime as an unscholarly or parochial organization. It is important that the AAG retain its reputation as an objective scientific organization that operates according to democratic principles. I hope the whole matter can be resolved quickly in a way acceptable to all.

From Carolyn McGovern, a graduate student at Syracuse University, to me:

> I for one am absolutely unwilling to return to the days of banned books and will not tolerate any censoring of my reading materials, particularly when they concern geography. At your request of June 4th, I have written the Executive Committee demanding the immediate resignation of Ms. Patricia McWethy for her irresponsible action concerning your book *The Immoral Landscape*, and for her outrageously presumptuous attitude in thinking that she has the right (and the position) to allow her emotional response to

a topic, as opposed to intellectual, deny others similar access to information concerning newly published material. It is painfully obvious that Ms. McWethy never understood the expression "Don't judge a book by its cover."

From Peter Gould, professor of geography at Pennsylvania State University and recipient of a meritorious award from the AAG, to me:

I think you are quite right—this really is quite extraordinary. Butterworths is a highly reputable academic publisher, and I cannot believe a request from them was denied. Unthinkable to my mind. I hope the members act quickly and decisively.

From James Allen, professor of geography at California State University at Northridge, to Pat McWethy:

I found it hard to believe that someone as generally well respected as yourself would set herself up as a censor, and I would guess that you have by now realized that your action was a serious mistake. If you have decided that you were in error and you have arranged to send the list to Butterworths or if there is some pertinent information that I should have on this matter, please let me know so that I will not need to carry the matter further.

From John Chappell, erstwhile geographer, to me:

I would probably disregard the flier if it came to me without comment. If it was juxtaposed with my own work I would be offended. I am rather puritanical personally, though left-wing politically. But it is absolutely intolerable that you should not have a chance to communicate your results with whoever [sic] is interested. To refuse you the chance, to refuse to let a social scientist write and communicate about a social malaise, is about as sensible as not letting a medical researcher write about a physical disease. If McWethy has never read your book, she surely would not know if there was a legitimate complaint against it—as if, for example, a doctor recommended smoking and other polluting, cancer-causing experiences. Yet, it seems her attitude, not yours, is immoral, as well as anti-intellectual and perhaps also illegal.

From Kingsley Haynes, director of the Center for Urban and Regional Studies at Indiana University, to Richard Morrill:

It appears to me that one person making these decisions for the organization is a dangerous precedent and that it is your responsibility as president to review this case with the council and make an appropriate organizational decision. Given the "junk" I receive via the AAG mailing list, I cannot see on what grounds an advertising flyer about this scholarly book could be seen as unacceptable. Topics that are simply controversial cannot be designated as out of the research realm of the field.

From Kevin Cox, professor of geography at Ohio State University, to Pat McWethy:

I earnestly hope that your decision on this is not final, and for several reasons. In the first place, it does seem to pose a threat to the freedom of expression which we all cherish. It could also set an unfortunate precedent in this area for future Executive Directors. You personally may be offended by the subject matter. Who knows what your successors might be offended by and, subsequent to your precedent, act on their indignation. In the second place, the profession can ill-afford the derision that will be heaped on us if the media become aware of this act of censorship—and surely they will since news of it seems to be traveling quite fast within the profession. In the light of these considerations, I sincerely hope that you will reconsider your position on this matter.

From Professors William Denevan, Robert Sack, Daniel Doeppers, and Dan Muhs, and librarian Miriam Kerndt (all at the University of Wisconsin at Madison), to Richard Morrill and John Adams:

If Richard Symanski's allegation is substantially correct, then we ask that you and the Council of the AAG appropriately reprimand Ms. McWethy and that you establish a clear policy which will prevent a recurrence. Censorship, or the appearance of censorship, cannot be tolerated in academic associations. The AAG mailing list should be available to any author or publisher of a scholarly book for promotional purposes regardless of the content of the book concerned.

From Carl Johannessen, professor of geography at the University of Oregon, to Pat McWethy:

I write now to urge that you not withhold the AAG mailing list from Butterworths Pub. Co. of Toronto. I am not requesting your resignation but I
am requesting that knowledge not be thwarted because a topic is considered unpopular or even objectionable to some. Since I have occasionally
been blocked from easy publication of innovative and iconoclastic ideas, I
feel the process deeply. Please reconsider your position on the basis of those
you represent rather than your personal view of how to control prostitution.

From John Agnew, associate professor of geography at Syracuse University and chairman of the social science program, to me:

Got your note. I'm concerned about the mailing labels policy too. From
what I've heard, and it's little, I'm not a big AAG favorite either. Make-it-
up-as-you-go-along has long been the process of "policy" formation at the
AAG. You just happen to have challenged the arbitrariness and you know
you're not too popular! What can I do to help?

Some geographers (I have no idea how many) were unhappy with my
protest. One past president of the AAG, John Fraser Hart, told me that he
thought my concern was legitimate, but that he would do nothing to help
me until I apologized to Pat McWethy for asking for her resignation. A few
sent anonymous letters. In one I was accused of "making a mountain out of
a mole hill—My Gawd!" I was told that rather than protest, I should have
simply gone to the library, found the membership list, duplicated it, and
sent it to Butterworths. "Any list that costs 2 cents per name really isn't
worth much," the anonymous critic reminded me. "Las Vegas casinos pay
$100.00 per name for THEIR lists; I once sold them one name and got a ck.
for $100.00. . . . I was stunned!"

I did not accept the decision of Adams and the AAG council. Instead, I
sought help from others, people still inside the academy (at this time, I did
not have an academic appointment). I asked Arthur Getis, then chairman
of the Department of Geography at the University of Illinois, if he would
call Adams and Morrill—both of whom he said he knew "quite well"—and
others. I asked him to plead the case that censorship served no one well,
that the longer it remained in effect, the more it would thicken the already
tattered image of geography. Getis would do what he could, he assured me.
But first, he said, he would have to consult with his colleagues to determine

whether this instance was indeed a case of censorship. It was during this conversation that Getis told me that he just couldn't wait to get his library copy of Gay Talese's *Thy Neighbor's Wife.*[8]

In mid-June, from Valley, Nebraska, I wrote a note to John Fraser Hart, a colleague of Adams at the University of Minnesota, a previous executive director of the AAG (the position now held by McWethy), and the editor of the *Annals* when my now infamous article on Nevada prostitution had been published. I asked Fraser Hart to reason with Adams and members of the council, most of whom he knew personally, and for at least a decade or more. I also wrote a short note to Adams, saying only that his decision and the decision of the council to stand by McWethy's censorship of my book were unacceptable.

Shortly thereafter, I drove a day out of my way to see Professor Risa Palm at the University of Colorado at Boulder. I knew that she had gotten her Ph.D. at Minnesota and was a good friend of Adams, and I sensed that she commanded a certain respect in the profession. (Before long, she would become a president of the AAG, in 1984, and later a dean at the University of Oregon). I petitioned her that day to call Adams and ask him to get the decision reversed. She said she would do "something." I never found out what she did, and would later conclude that she didn't do anything.

I thought that by the time I returned to Illinois in late summer, the censorship issue would have been resolved. But instead, upon my arrival home, I came upon a piece on the front page of the August 1982 issue of the *AAG Newsletter* (which is sent to all members of the AAG). It was an announcement meant to quiet me and the dissenting voices of geographers who had stood behind me. In this note, the AAG membership was supplied with an IRS Code Section, and reminded that the code says that "excessive income from such sources [the sale of mailing labels] could jeopardize our tax-exempt status." Furthermore, it noted that the AAG "cannot be concerned with promoting the interests or personal gain of individual *nonmembers* and their publishers. AAG membership labels have been sold on occasion to publishers if it is clear that no member could feel that the mailing was inappropriate for the AAG to sponsor. The gains from the sale [of mailing labels] are minimal and not worth the possibility of alienating a *single* member" (italics added). A new, and more perverse, angle had been introduced: it was not just that the rights of a minority were more important than the rights of the majority in this First Amendment issue, but that the minority was a "single member."

On September 1, 1982, I received a letter from Morrill in which he reiterated much of what Adams had said to me in his letter of June 10. To Morrill, the "affair was one of those recurring real life events for which there is no satisfactory resolution." He went on to inform me that "over the years the Executive Director became extremely sensitive to events which lead to membership disaffection and decline. In her judgment, and according to her duties, [Pat McWethy] felt that the AAG would be liable to severe reaction, even litigation, from members who objected to the release of their names for promoting your book." Thus, it seemed, Morrill confirmed what had been said in the earlier newsletter, namely, the AAG was run by minority rule. If a few did not want to receive the flier on *The Immoral Landscape,* then it was more important than abridgment of the First Amendment.

In September 1982, a graduate student at Syracuse University wrote to me and said that she had been threatened with a lawsuit by McWethy because she had let McWethy know in quite compelling terms that she abhorred censorship. By this time, Butterworths in Toronto had gotten calls and letters from McWethy, informing them that the AAG could sell their mailing labels to "whomever it so desired," and, furthermore, that this affair with *The Immoral Landscape* was hurting the relationship between the AAG and Butterworths. With this turn of events, Ernest Hunter, president of Butterworths in Toronto, decided that my book was not worth the trouble it was causing him. He called me and said that he was not interested in pouring money into a First Amendment legal battle. He wrote to McWethy expressing his regret over the whole matter. Hunter made it clear to me that thenceforth the fight was mine. To McWethy, he wrote: "[O]bviously I can not speak for Professor Symanski, our author, though I have no doubt he can and will present his own case. It might appear that Butterworths is in the middle of this dispute, but it is neither our intention nor desire to be a participant. As far as we are concerned, this particular correspondence is closed."

There would be no more efforts made by Butterworths to buy anyone's mailing list. There would be no more money spent on advertising. And the only question that now remained at Butterworths was how to get rid of all the freshly minted copies of *The Immoral Landscape* in the warehouse. Butterworths informed me that I could have the remaining copies of *The Immoral Landscape,* virtually the entire run of two thousand copies, for nine hundred dollars, plus whatever it would cost me to pick up the books at the Toronto warehouse.

I now called Professor Greg Knight at Penn State University and complained about being ripped off by the AAG. "What right," I asked, "do you guys have to take breakfast at the Sheraton with my royalties while you won't even give me a cup of coffee for my generosity?" I reminded Knight that he was editor of the AAG series, Resource Publications in Geography, when I and John Agnew, then at Syracuse University (now a professor and chair of geography at UCLA), signed over all royalty rights to the AAG for *Order and Skepticism*, a monograph that we had written. I said it was now easy to conclude that the AAG, among other things, had a double standard, something to the effect that: "We'll take what you write and take the money from it too" (by the end of 1982, the AAG was making profits on *Order and Skepticism*), "but we won't allow your publisher to buy a mailing list from us for a book you have written about prostitutes." I reminded him that I had not been a member of the AAG when I wrote or he accepted *Order and Skepticism*, but now, suddenly, my not being a member of the AAG had become a major pretext for not selling the mailing list to Butterworths.

Knight said that he saw the contradiction. He added that he did not agree with the censorship of *The Immoral Landscape*, and he asked what he could do to rectify matters. We talked about various options, and then he suggested a three-way phone hookup: me, himself, and the next president of the AAG, his colleague at Penn State, Professor Peirce Lewis. They would call me soon, he said. The call never came. Nor did I receive so much as a short note from either Greg Knight or Peirce Lewis.

Greg Knight had told me that Salvatore Natoli, the AAG's educational affairs director, had control over the sales of *Order and Skepticism*. I wrote to Natoli in October 1982 and said that I was grossly offended by the censorship of my book and the subsequent cover-up. "Don't," I said, "sell *Order and Skepticism* to anyone until the censorship is lifted and Butterworths gets the mailing labels. Don't sell it until the *Newsletter* cover-up is retracted and the guilty parties identified." I further told Natoli that I would much prefer that the remaining copies of *Order and Skepticism* be burned if the censorship was not lifted. As the senior author of the monograph, I assured Natoli that he did not need John Agnew's permission to honor my demand. And in any event, I said that I was certain that Agnew would concur with my decision.

It took Natoli six months and another angry letter from me before he wrote to me. He said that Greg Knight was mistaken, that it was only the AAG council that had the authority to stop the sales of *Order and Skepti-*

cism. Natoli added the following in his March 1983 letter: "I feel that the ban on the sale of *Order and Skepticism* might appear to be considered censorship or even restraint of trade moreso [sic] than on your book, *The Immoral Landscape.*"

Soon thereafter, Natoli sent me another letter, and in this one he said that my demand to stop sales of *Order and Skepticism* would be weak, "given the fact that your past membership record has been spotty."

I did not go to the trouble to inform Natoli that no one asked me if I was a member of the AAG when I wrote my share of *Order and Skepticism* and assigned all royalties from the monograph to the AAG. I had not, in fact, been a member of the AAG since the mid-1970s. I did not begin writing *Order and Skepticism* until 1979.

One month prior to the October 1982 AAG council meeting, I talked with a lawyer in Champaign, Illinois, who specialized in First Amendment rights, and, as coincidence would have it, the First Amendment rights of prostitutes. He examined the facts and evidence of the case that I put before him, and he concluded that I had a clear case of conspiracy censorship under Section 1888 of the Civil Rights Code. In his opinion, I had a better than four-in-five chance of winning a lawsuit against the AAG. I brought this information to the attention of Professors Adams, Morrill, Lewis, and several others. I did not explicitly say that I would file a lawsuit against the AAG.

In late January 1983, I heard from Susan Hanson, some years later, in 1990, president of the AAG, now a member of the National Academy of Sciences. I had asked her what her position had been on the censorship matter. To her, it was all a matter of strategy, not a First Amendment issue. She wrote, "You ask about my role in the 'release of the mailing list' controversy. Like several others I know, I called Pat McWethy and suggested that the best strategy was the one that was in fact adopted by the Council at its October meeting: allow members to say whether or not they want their names released to agencies or firms requesting the AAG mailing list." But had the censorship of *The Immoral Landscape* been lifted? Susan Hanson did not say. Did the council even consider it a case of censorship? She would not say. She clearly did not want to be implicated in the net I was throwing around the discipline's spineless and immoral power structure.

More time passed, and I heard from no one about what happened in that October AAG council meeting. Then in the middle of February, I wrote a note to Peirce Lewis, now the vice president of the AAG. I said that if a just answer was not forthcoming soon, and the censorship matter settled in my favor, his year as president was going to be an uncomfortable one. I did not elaborate.

Lewis wrote to me at the end of February. In his letter, he said:

> At the October meeting of the AAG Council in San Diego, the Council okayed a resolution which will make available the Association's mailing list to anybody who wants to buy the list to market scholarly material. The exact wording of the resolution is reproduced in the February 1, 1983, *AAG Newsletter*, p. 13 [he did not enclose a copy]. Individual members of the AAG will have the option of having their names removed from the list if they so request. As far as I am concerned, this action should satisfy those who want the AAG list used for general dissemination of scholarly information, and will simultaneously protect the privacy of people who desire such protection.

It would be another month before I would get a copy of the *AAG Newsletter* report to which Lewis referred in the opening paragraph of his letter. The search for rationalizations had not ceased. No mention was made of what had caused the furor to get the council to come forth with the resolution. There was, however, an attempt to justify why the mailing labels had not been sold to Butterworths. Butterworths's problem was that it had wanted the mailing labels to advertise only *my* book. "Council members were provided with a list of entities to which mailing labels had been sent in 1981 and 1982. Some 36 sets of labels had been sold, including 20 full membership lists, of which 7 were to commercial publishers. In no case was a list sold to a commercial publisher for solicitation of offers for a single book." Presumably, I was to read, if there had been another book on prostitution that Butterworths was selling at the same time, the AAG would not have engaged in censorship!

Lewis had more of relevance to say in his letter to me. "In my opinion," he wrote, "the Council discussion and vote removed any obstacle to Butterworths buying the list to advertise your book." He apologized for the fact that no one had informed me or Butterworths.

Because, in Lewis's opinion, the AAG had now finally righted the wrong, he wanted to assure me that the censorship would have been taken care of straightaway if only I had sent a courteous letter "to the Central Office to reconsider its denial of the list to Butterworths." I understood him to mean that the prolongation of the censorship of my book, and presumably

all of the rationalizations behind it, would never had occurred had I not expressed indignation over the censorship of a book on which I had spent five
years of my life.

Peirce Lewis, who would soon be president of the Association of American Geographers, ended his letter to me by congratulating himself and his
censuring colleagues. "It is, in fact, a *considerable tribute to the AAG
Council that Butterworths' request received a fair-minded hearing*" (italics added).

In the spring of 1985, I attended the Annual Meeting of the Association of
American Geographers in Detroit. I sought John Adams. It was the first
time that I had talked to him since he had written the letter of June 10,
1982, backing McWethy's decision to censure the book. I asked him why he
had gone along with the censorship. He said, "I had nothing to do with it. It
was a committee decision."

That year, an entire session of the annual meeting was devoted to the
issue of censorship and freedom of expression. Fewer than twenty people attended. In addition to a guest speaker from the American Association for the
Advancement of Science, three geographers sat on the free-speech panel:
Professor Risa Palm, who was now president of the AAG; Susan Hanson,
now editor of the *Annals;* and Wilbur Zelinsky, 1972 president of the AAG,
meritorious award recipient of the AAG, and for more than two decades one
of academic geography's most vocal and visible champions of oppressed minorities, especially women. It was the unchallenged consensus of the geographers on the panel that there had *never* been a case of censorship in the
history of the discipline. They wanted it to be a matter of record that geographers were ever vigilant to suppression of information and matters of free
speech and would quickly rally to fight off any hint of censorship. I attempted to challenge this bald lie with a question. I was ignored.

As Zelinsky left the meeting hall, I stopped him and asked him why no
mention had been made of the censorship of my book. He gave me a puzzled, angry look, and then said that he did not know what I was talking
about. I tried to remind him of what had happened to *The Immoral Landscape*, but in midsentence he scoffed and threw up a hand and walked away.

The AAG rationalizations for censorship that were thrown at me, and others, deserve some elaboration. Selling the mailing list to Butterworths
would have brought in $125, which, according to John Adams, would have
jeopardized the tax-exempt status of the AAG. The fact is that the $125 re-

ceived from Butterworths would have been less than one-half of 1 percent of the AAG's gross annual income. No IRS agent or CPA I have ever known (into the 1970s, I was licensed as a CPA in California and had done both tax consulting and audits on nonprofit organizations) would have considered making a case against the AAG or any other tax-exempt organization unless the amount was on the order of a magnitude greater than $125. Assuming, for purposes of argument, that the *total income* from the sale of the mailing list was indeed "material" (the defining concept used by accountants), the tax-exempt status of the AAG might then have been challenged. But were it to occur, it would come about because of the sale of the mailing list to hundreds of publishers and others, not because of the sale to any specific publisher or a couple of dozen of them for purposes of promoting books on the geography of prostitution, or of child pornography, or whatever one might imagine worthy of book-length treatment.

Another pretext for censuring the book was, according to Adams, that there had to be "compelling reasons from a 'services to membership' standpoint." But surely this criterion is ad hoc, and one, that in any event, could not have been met if put to the test. The AAG has long been in the practice of routinely selling its mailing list to all kinds of publishers—and others— many of whom peddle marginally scholarly books or books without explicit geographic content or services in any way related to the mission of the AAG.

It struck me then, and it strikes me now these many years later, that to lump academic publishers, and especially ones with a long and distinguished history such as Butterworths, with airlines, travel bureaus, and rental car and insurance companies is an outrageous insult. The service performed by selling a mailing list to academic publishers is clear. As Adams noted early on, it is no small puzzle trying to decide what these other institutions have to do with geography or the pursuit of truth. If it is these kinds of institutional fliers that an academic membership rails against when it objects to the mailing list being used for advertising, then it has a right to object. If AAG members did not want to receive fliers on academic books, however, then they should have registered their names to that effect with the AAG head office, or simply thrown away the "junk mail" upon receipt, something all of us do daily. If, on the other hand, a few members were opposed to receiving any fliers, and their minority voices were to be heeded at the expense of the majority—which clearly does want to know of new books—then Adams and anyone else who would put a minority interest before that of the majority were violating a central tenet of democratic rule.

It likewise seems patently untenable to have argued, as Morrill did, that there was reason to fear that the AAG would be "liable to severe reaction, even litigation, from members who objected to the release of their names for promoting your book." McWethy or Morrill or others who might have taken this position may have had a point, for example, in the case of a picture book of child pornography where there was no pretense of scholarship. But *The Immoral Landscape* is clearly not about pornography, and it is documented to the point of embarrassment. And insofar as the book contains words that some would consider offensive or in bad taste, these words, are, without exception, parts of direct quotes, taken from either the diaries of prostitutes, secondary sources, or street or brothel interviews conducted by me. In addition, not a single charge of any of these sorts was ever made by McWethy, Adams, Morrill, Lewis, or anyone else in the AAG. Indeed, I have no evidence that anyone directly involved in the censorship had even seen a copy of the book at the time that Adams's letter of June 10, 1982, stated that the Executive Council of the Association of American Geographers had backed McWethy's decision.

Pat McWethy made a mistake, but at least she might have offered the excuse that she did not have research credentials: she had no Ph.D. and had never done any legitimate research. She might have pleaded the erstwhile feminist logic that an open discussion of prostitutes hurts women and therefore ought to be suppressed—which she did not do, and which in any event is not a tenable reason for censorship. She might have pleaded that she was doing what her bosses had instructed. But Adams, Morrill, and all who were party to Adams's letter had no such excuse. They had at their disposal the presumed "good sense" of a collective mind: a committee. They had a chance to champion and reaffirm an inalienable right, one of the unmistakable cornerstones of the United States Constitution. And like all university professors, they were, and are, almost daily, reminded of the sanctity of the principle of academic freedom.

It was astonishing to me at the time, in 1985, and it remains equally astonishing to me to this day, that two people who were intimately aware of the circumstances surrounding the censorship—Risa Palm and Susan Hanson—and a third, Wilbur Zelinsky, who undoubtedly knew of the controversy, could so blatantly assert in a public AAG forum, one including a representative of the American Association for the Advancement of Science, that there had never been a case of censorship in the history of the AAG. At the very least, all of these individuals owed their audience an explanation as to why, in their opinion, the refusal of the AAG to sell its mailing list to Butterworths for purposes of advertising *The Immoral Land-*

scape did not constitute censorship—if that's what they believed. The book, the controversy, my charges: nothing related to it were so much as hinted at during the free-speech panel held at the 1985 Annual Meeting of the Association of American Geographers.

To this day, I do not have "inside" information that would explain why the ruling elite of the Association of American Geographers would have engaged in a blatant case of censorship. I can only speculate that several factors combined to result in a legal conspiracy, and a prolongation of that conspiracy—one that very much contributed to my being blackballed for life from academic geography.

McWethy had clearly erred, and it may well have been apparent to some on the AAG's national council at the time. But once the error became "national news" within the profession via my widely disseminated letter asking for McWethy's resignation, the council may have seen the issue as that of either me or her. Notwithstanding the obvious First Amendment and academic freedom issues, it might well have seemed easier at the time to side with McWethy. She held an important and highly visible position in the discipline, was well known to Adams and the members of the council, and was in constant communication with them. I, by contrast, was an outsider. At the time, I held no academic position, and I had a blemished reputation that was now picking up steam because I had fought the censorship. For several years, a number of people in the profession had been upset with me for two articles I had published in the *Annals* in the mid-1970s, the 1974 one on prostitution in Nevada and the 1976 one on the use of language by the historical geographer D. W. Meinig.⁹ My image was further blackened by the belief that in 1976 I had been fired from a tenure-track position in the geography department at the University of Texas at Austin. That this rumor was patently false, that a promotions committee in the College of Arts and Sciences had voted unanimously to overturn a departmental decision to not renew my contract, was largely irrelevant. I was carrying a sufficient amount of "negatives" to make another negative story about me seem credible. In this light, it might well have been thought—or imagined—that any gripe coming from me could be discounted if not ignored; credibility was clearly with the power structure of the AAG, not with me.

Not to be discounted, of course, is the subject of the book itself. Until the last decade or so of the past century, inquiry into sex topics has never been held in high repute in the academy, even in sociology, which has a fairly long tradition of such inquiries. All claims about liberal leanings to the contrary, the academy continues to be pretty much a reflection of the larger society within which it is imbedded: staunchly conservative on matters related to sex, often reactionary, and quite frequently judgmental in morally queered ways, irrespective of whether an evidential base has been considered. Up to the time *The Immoral Landscape* was published, there were fewer than a handful of Ph.D. dissertations on prostitution in the previous one hundred years in the social sciences. Prostitutes and people who would associate with them in order to get information about their world have been seen as immoral, reproachable, and outside the law.

And so McWethy had—still has, no doubt—her backers, and for precisely the reason that she noted: nay, not just that some people were (and are) offended by the subject matter, but rather that *many* people were (and are) so offended. A measure of this sentiment, I think, is the extent to which *The Immoral Landscape* has been ignored by virtually everyone in geography who has written about feminism, or female issues—male and female alike. And this disregard is admittedly more than a little strange, given how prominent prostitutes are in major cities, how they tax city budgets, how central they are in the drug culture and in spreading the HIV (much more so in the Third World than in the industrialized world), and the extent to which, in *The Immoral Landscape,* I gave attention to injustices and unfair treatment of women before the law.

To this day, then, it is as if *The Immoral Landscape* does not exist. Except to this or that offbeat geographer trying to make a name for himself by making claims for a new cultural geography, prostitution remains a nonissue, as hidden from view as most societies have long tried to keep the institution and its practitioners.[10] To be sure, the case here is more general. And as with American society as a whole, geographers—alas, American academics—have not done a very good job of getting out of their Victorian corsets.

It is quite possible, as so often happens in committee meetings, that one or two powerful individuals pushed the case for backing McWethy, and no one had the will or courage to counter. Power and the ability to foresee the consequence of one's actions bear no necessary relationship to one another. Junior or nonvocal members of the executive council may not have cared enough about the issues, as compared to how much they cared about their images or growing power bases, to have thought the risk of confronting Adams and McWethy or anyone else centrally involved was worth the personal cost. As members of a committee, they could disappear, as Adams did, behind a "committee decision." It was everyone's decision; it was not mine. Thinking thus, the perceived cost to any given individual for what later might be seen as a wrong or stupid collective decision would be minimal, less than the perceived cost of "bucking" the system, or the one or two more powerful individuals guiding the decision.

It is conceivable that, notwithstanding the widely accepted principle of academic freedom, Adams and others on the executive council had only a vague sense of what the principle in fact means, or none at all. I said that Richard Morrill had claimed that "neither [he] nor the AAG engaged in censorship in any form"; it did not "because the book was already published and sent out for review by A. A. G. publications." To Morrill, who was inextricably involved in the censorship, the "AAG action" was nothing more than "an inconvenience" to me, and "would not fundamentally affect [my] freedom of expression or pursuit of [my] profession." This peculiar kind of depraved reasoning was coming from someone who, in a presidential address before the Association of American Geographers, published less than two years after the censorship of *The Immoral Landscape,* wrote that geographers have a special responsibility to safeguard truth, that they must be ever vigilant "against attacks on freedom of inquiry and expression."[11] Perhaps for people like Morrill, academic freedom means the right *for them, as individuals,* to write and speak within the university as *they* so wish; it does not mean that it is a right that extends to others. And insofar as it does, abridgment of *their* freedoms of expression is no more than an "inconvenience" and therefore nothing that anyone would care to defend or fight for.

Twisted and Running Scared

> He that will not reason is a bigot; he that cannot reason is a fool; and he that dares not reason is a slave.
>
> —Sir William Drummond

I made three vigorous efforts to get the *Annals of the Association of American Geographers* to publish "An Anatomy of Censorship." When Susan Hanson was editor of the *Annals*, from 1985 to 1987, she refused to do so. When Stanley Brunn was editor of the *Annals*, from 1988 to 1993, he refused to do so. And when Carville Earle was editor of the *Annals*, from 1994 to 1996, he refused to do so. In every case, I made it clear I was willing to go to almost any length to have it published, but not one of these editors had anything like the courage or vision required. All were, and are, very much an integral part of the venal and effete geography establishment, utterly unwilling to confront a dark chapter in the history of the AAG.

As the story of my third effort to publish "An Anatomy of Censorship" in the *Annals* was unfolding, I was contacted by the editor in chief of a major academic publishing house who said he was eager to republish *The Immoral Landscape.* He said that the book was now a "classic," and that it needed to be in print. After some persuasion by him, I agreed to do a minor update, but only on the condition that he publish—unreviewed—an appendix called "An Anatomy of Censorship." He agreed. For almost three years, I fitfully worked on the revision of the book, virtually the sole motivation being that appendix that I wanted to see in print. But then one day, for a variety of reasons, I lost interest in the update of *The Immoral Landscape* and returned the advance to the publisher.

The final effort to get "An Anatomy of Censorship" published in the only journal where it really belonged, the *Annals of the Association of Ameri-*

196

can Geographers, was made in 1994. I submitted the article—almost identical to what is herein published—to Carville Earle, the editor, on January 19, 1994. It would not be until November 9, 1994, that I received a final letter of rejection. In between, several letters were exchanged between Carville and me, and there were also telephone calls that I made in an effort to do whatever I could to reason with Carville, to get fair reviews, and to get the piece published.

A summary of some of this correspondence, a personal journal entry, and one of my letters to Carville in its entirety reveal the extent to which the last of the discipline's editors involved with this fiasco was just plain scared of the narrative and the consequences were he to publish it.

In the submission of the manuscript to Carville Earle, I acknowledged that I was "putting [him] between a rock and a very hard place," and that not only would some quite high-profile people in the discipline be upset by the article were it to appear in the *Annals,* but their reputations would also probably be damaged by the revelations. I noted that its publication would be also good for the discipline, the demonstration of a willingness to confront an ugly truth. In the cover letter, I suggested a number of ways in which Carville could, as I put it, "protect his ass." Rather than get two or three reviews, I suggested that he get between six and eight. I suggested he get responses from the people I accused of bad behavior and publish their responses along with the article. And I also suggested that he consider including a note on his rationale for publishing the piece.

On February 8, Carville wrote to me and said that he had read "An Anatomy of Censorship" with "unusual, even prurient interest." He added that "the prospect of a discussion of malfeasance in high offices of the AAG is all too irresistible." But then in the next line, he said that the narrative was appropriate for the Op-Ed section of the *Chronicle of Higher Education,* "or perhaps a journal of administrative law." The narrative, he said, was not "the usual stuff of a scholarly journal committed to advancing geographical scholarship." He also claimed that "the controversy is several steps removed from the *Annals.*" And with this illogical turn, he concluded that he would not consider publishing "An Anatomy of Censorship." He had not sent the article out for peer review.

Four days later, on February 12, I wrote to him and accused him of engaging in the very kind of censorship that the AAG had embraced fourteen years earlier. And I took note of his spurious reasons for doing so. I urged him to get the paper reviewed, and to get signed reviews to avoid "cowardly and cheap potshots, to force people to be accountable for their charges and the quality of their reviews." I heard nothing from Carville for almost six months. On August 5, 1994, I called him to ask if he'd gotten reviews of the article. My unedited personal journal entry of August 5 reads as follows:

> Called Carville. I'm getting impatient. Damn editors, they're all pretty much alike. They take forever and then they're always full of bull-shit excuses. They promise they'll get a ms. back to you in two to three months and it's often double this time. And then the reviews are, with a rare exception, not reviews at all.
>
> Carville said he'd sent out reminders to the first three reviewers he sent the ms. to and finally got a review from one of them, but not the other two. Then he chose three more reviewers, got no response, so he sent out reminders. Again he only got one review. Carville assured me he'd soon make a decision, even if no further reviews were received.
>
> Then Carville shifted gears, seemed eager to tell me something that he said that I, of all people, might understand. Recently, he said, he'd received an unsigned missive on UCLA School of Architecture and Planning letter-head calling him a "fucking asshole in a fourth-rate university," "with his brains in his fucking ass for putting abstracts at the end of articles," and "not being nearly bright enough to be the editor of the journal." There was more, including accusations that he had a duty to publish articles on the oppressed of the world and "fucked over women." Carville said he didn't know what to do about this unsigned missive. He thought of publishing it "as is" in the journal. But he decided against this, and is considering bring-ing it to the attention of the Executive Director of the AAG. I told him this'd be a waste of his time, that the Director's another disciplinary light-weight, without tendons in his knees. I asked him if he had any good and reliable friends at UCLA. He said he thought maybe one, but he wasn't sure. I finally told him, Forget it. I've gotten several of these kinds of un-signed nasty missives over the years from academic geographers. Fucking cowards—all of them, I told him. He went on: that the job didn't pay him enough to have to take this kind of shit, that he didn't really care that much about the job. He said he was finding it hard to put the quack letter in a box, kick it a good one, and then burn it. He was clearly rattled.

September 27, 1994

Carville Earle, Editor
Annals of the Association of American Geographers
Department of Geography and Anthropology

Louisiana State University
231 Howe/Russell Geoscience Complex
Baton Rouge, Louisiana 70803–4105

Dear Carville:

I have finally received your letter of September 16 informing me that you have "[sent] along the two reviews" [*sic*] you have received on "An Anatomy of Censorship." I wish to protest in the strongest form possible the conclusions you have reached, and the audacity of sending me such shabby and unconscionable reviews. In order to anticipate that I will have to circulate this letter widely to influence your approach to this manuscript and its ultimate disposition—and will do so primarily through cyberspace—I will quote in full your paragraph to me summarizing the "review" from the reviewer whom you said "explicitly stated that [you] should not send [me] a copy of his review," as well as the one review received and sent. The summary of the purported review first.

Carville Earle's Summary of Review Not Sent to Me

As for the other reviewer, I think that I can say that his/her unease stems from three concerns: 1) the intensely personal nature of the statement; 2) a tone that reads more like a legal brief (and hence an adversarial tone); and (ergo), 3) the inappropriateness of the *Annals* as an outlet for an essay (and an issue) that would be better served by a forum that includes legal experts.

The One Complete Review

The writer should know that *I* did *not* think it "dispassionately" written . . . (p. 34). Anyway, I am *pleased* to have received this submission—since I have long had a copy of *The Immoral Landscape* (1981), and thought much of it to be intriguing and geographically interesting when I read it long ago. I awaited the verdicts of clever reviewers. . . .

My incomplete responses above should suggest I am *not* "informed" on the "topic" of the paper. My "rejection" of the paper would be on the grounds that it is not a balanced treatment of the incident, neither is it well-composed (or as compellingly written as it might be). And it is journalistic, rather than scholarly; pity . . .

Perhaps Richard Symanski should consider this question—if he could find a disinterested person (probably a scholar; not necessarily a geographer) able to analyze and synthesize objectively and compose really compelling narrative, would it not serve readers of the *Annals better* if the transcripts—letters (etc., etc.) were turned over to this person, who could then pursue his/her own enquiries and, in due course, produce a perceptive paper that might inform, interest and stimulate those readers of the *Annals* with an appetite for such pieces? I would envisage such a piece having "commentaries" from Symanski and others brought into the story (if they wished to comment).

I would be concerned if you published this that so many persons are "quoted" (some in an off-hand way) who are not able to respond—or may not wish to be "quoted."

Imagine, Carville, what you as an historical geographer of some distinction are saying about your own standards: you have none! You, even quicker than me (since I am not a historian) should have thrown out the one review received (hereafter #2) as soon as you read the very first sentence. Not a shred of evidence (by way of citing an adjective, an adverb, a phrase, a sentence, an idea) establishes that I am "*not* dispassionate" (underlining not mine), that the paper is "not balanced," that it is "[not] well-composed," that it is "journalistic, rather than scholarly," that it is not "objective," that it is not a "really compelling narrative," that it is not "perceptive." I mean, really, Carville, even for someone who loves fiction as much as I do, and has written one nonfiction book without a single footnote or bibliography *(Wild Horses and Sacred Cows)* and another without footnotes *(Outback Rambling)*, I would not dream of making charges of the sort made repeatedly by this reviewer without, at a minimum, citing one or two examples from the text for each of the charges. (After all, there is a purpose to using something called examples by way of illustrating the general case.) That you would accept such charges against me without even minimal documentation is, quite frankly, more an indictment of you than this shabby second-rate community college reviewer. To sound like a father talking to a wayward son who clearly knows better: Shame on you, Carville!

I assert that the paper is quite dispassionate (very unlike this missive soon to be shooting through cyberspace), and to the extent that I am wrong, please cite page, line and word, and if I am guilty as charged be assured that the word or the line or the thought will be excised faster than you can bring that ever-present cancer stick to your mouth.

But oddly enough, you apparently do not see the contradiction in your own letter, to use your own words, the paper has "a tone that reads more like a legal brief. . ." Well, of course it does, and by quite conscious design! Nor, apparently, can you see that if the paper indeed reads like a legal brief, it is not likely to read at all like "journalism," as reviewer #2 asserts but does not begin to establish.

Reviewer #2 states that he is "not informed on the 'topic.' " Then, pray tell, what right does he have to review the manuscript! He has in effect said: get someone more competent, don't listen to me. Well, why didn't you—or, to save me time, throw out the review as a responsible editor would—or should—do?

In the second paragraph of this review, there is, I must say, a small tickling jewel. The manuscript is not "well-composed (or as compellingly written as it might be)." Well, well, love of words and love for the English language that I have, and pride that I take in much of my prose, I would love to cross prose swords with this mighty critic—if only he (or she?) would have the courage to come out from behind the bushes and let one and all look at his or her compelling prose. But, then again, and who knows, maybe this manuscript on censorship is one of my undistinguished efforts. But if it is, at least do me the courtesy of giving me page, line, and so on where my prose drops below the exalted standards of the *Annals* rather than hanging me here as elsewhere with a sailor's noose without even a pretense of a show of evidence. But then maybe this reviewer, like you Carville for accepting this drivel as a review, is an ardent subscriber to esteemed supermarket rags like the *Enquirer* and the *Star*. (I mean, four million people a week can't be wrong in swallowing the undeniable truth that Elvis is alive and well in Flint, Michigan!)

Now to another matter, reviewer #2 wanting me to turn over all my materials to someone who is "objective" and can "synthesize" and "compose really compelling narrative," and, of course, "produce a perceptive paper that might inform, interest and stimulate. . ." My, my, my—some antidevulian sort who actually believes that objectivity is possible! (Little did I know just how postmodern I really am; and, gads, I sure hope Michael Dear doesn't find out.) But again (and again and again and again ad nauseam)—where is the EVIDENCE that I have not been "objective," etc.? Dammit, I do roundly object to being shot and quartered and salted for storage without due process, a decent hearing.

Our distinguished reviewer who so blithely ignores evidence, and swimmingly embraces kangaroo courts and cheap shots, is exercised that some dead minor players in the censorship story like my former Texas colleague George Hoffman (who told me on more than once occasion that I should not publish anything having to do with prostitution) cannot respond to my charge that he said this. Well, not only is this irrelevant, but knowing George as I once did (always quick to barge into my office from his adjacent office to authoritatively update me on whether it was now publication #183 or #184 or whatever that he had just added to his one-of-a-kind CV [while, it should be noted, rudely pushing aside the student in my office and blowing cigar smoke in her pretty face]), he would, let me assure you, be most eager to confirm that he not only said what I said he said in this regard, but would probably quickly add—even from the grave—that he was positively embarrassed that I had the audacity to talk to and write about women who sold access to their bodies because they were sick beyond words of making minimum wage and being exploited at Jack-in-the-Box and thereby starving to death or forever living on the sordid unfriendly edge. Something similar might be said about that equally distinguished one-of-a-kind geographer, Joe Spencer, who, as editor of the *Annals*, was supremely notorious for rejecting manuscripts out of hand because they had a few numbers or equations in them, and then, in all of his mighty editorial wisdom, somehow managed to produce some of the most wooden and unreadable and prolix manuscripts ever to grace the pages of the *Annals*.

It also offends our reviewer that some of the putatively still alive people I quote in "An Anatomy of Censorship" may not wish to be quoted! Well, I don't doubt this for a moment if they finally woke up to the fact that they were party to a blatant case of censorship. But, of course, they can in fact object in print, right down to contesting that I quoted them correctly, even calling me a liar if they so believe that I wronged them.

Now for the summary which you said distills the essence of a review that you may have received, a summary that I categorically reject as constituting a review. If I had full control of my faculties at the moment, I would not stoop to give this summary any words at all, but in my own irrepressible way I cannot but laugh at the complaint that, to use your words, the manuscript is "intensely personal." Well, it certainly ought to be! Alas, too bad the charge is not really true, and that the manuscript, as noted in the second complaint, is, in point of fact, so legalistic (and boring) in tone. By design, to reiterate. But it is the third complaint from your cowardly reviewer that most tickles, namely that the paper is best suited to "a forum that includes legal experts." My, my, my—it was perfectly okay for the Association of American Geographers to censor me without first seeking legal advice, but now you cannot publish a narrative of bad behavior by those running the AAG because the discipline is not full of "legal experts." Wow! The reasoning abilities of those who inhabit the profession of geography, and particularly the upper staircase that reaches toward Heaven, never cease to amaze me. I mean, I do not see why a story about the demise of Harvard geography (which the *Annals* published) should not have been submitted to a forum of psychological or psychoanalytic experts, or why my paper on feral horses in outback Australia, which you're publishing, should not have gone to a panel populated by outback arid land conservationists who wear long suspenders.

Okay, enough said to convey the message that I find the review and the summary and, worse, your acceptance of them, an insult to my intelligence. "An absolute and utter outrage," my biologist wife, Nancy Burley, said today after reading the review and your letter. "Simply shameful." (And lest you think she's an outrageous Alpha like me, let me assure you that she's about as demure in the old-fashioned Southern sense as women come in the 90s.)

If this is the best you can do on getting reviews for this or any other paper, then I heartily suggest that the profession would be a lot better off if the *Annals* had no reviewers whatsoever and instead merely sought an editor of considerable intelligence and independence who made unilateral decisions about what appeared in the journal. Put differently, I am simply numb at the realization that: (1) you did not have the good sense to simply throw out review #2 on the quite simple and straightforward rationale that it is (a) no review at all, (b) irresponsible and dead wrong in its charges and this should be transparently obvious to you since you have told me that you have read the manuscript, and (c) a mockery of the review process and a visible stain on the discipline to send this kind of a review to anyone in the discipline.

Let me further note that I am aghast that someone of your apparent talents, as displayed in your book of historical essays published by Stanford University Press, would have such utter disregard for evidence, and demonstration. I find it very hard to believe that you would tolerate such disregard for evidence in even a minor essay on some minor historical/geographical event in the pre-Lincoln South. For you to tell me that you consider these quite legitimate reviews leads me to only two conclusions: either you are as running scared as are all of those who ran from reviewing this manuscript (the hypothesis I want to favor), or, alas, I have greatly overestimated your good sense and sense of fair play and must therefore conclude that you are simply not good enough to be the editor of a journal, major or minor.

Indeed, if others are receiving reviews and the quality of reasoning you have displayed with me with regard to this manuscript, then I strongly urge you to resign as editor of the *Annals*. And I hope that others send you the same message.

Just before finishing this letter, I once again talked at great length with you on the phone, and I heard the following. That you do not yet have the courage to make a decision on the manuscript, but are strongly leaning against publishing it; and that you may or may not seek a third review but the likelihood of getting a "truly superb" review that would make you go with publishing the piece is slim at best. That you have not the least bit of interest in my plea that it will be in everyone's best interest, especially the profession's, for you to maintain control of the manuscript by publishing it. You really don't want me to rewrite it and put passion into it and then be my own editor, do you?

Well, guess it's time to get home to watch my boy Cole hit a few homes runs over the fence. Come tomorrow I'll get out my e-mail address book, get other addresses from the Internet, and then begin pushing magic buttons, urging all who read this letter to reason with you in a way that might be more effective than hour-long phone calls and impassioned pleas for you not to repeat the same mistakes made by Pat McWethy and John Adams and Richard Morrill and the entire 1981–82 AAG Elective Council.

On November 9, 1994, Carville sent me a letter saying that he'd rejected the manuscript. I immediately sent him a letter, again telling him he didn't have what it takes to be an editor and that he should resign. And I reminded him not only that I considered him incompetent but also that his bag of "Editor's Dirty Tricks was just a continuation of the cover-up of the censorship."

On November 21, 1994, I sent him yet one more letter, and in this one I drew his attention to a critical fact that I had completely overlooked in my narrative. The pertinent section of the letter reads as follows:

> This afternoon, while my son Cole took a nap, I opened some musty files, full of field notes and yellow reprints and colorful maps of famous streetwalking areas in major European cities. As I scanned the materials and wondered what might be useful in an update of *The Immoral Landscape*, my eyes fell upon a letter addressed to Lee H. Bowker, Director,

Center of Advanced Studies, School of Social Welfare, P. O. Box 786, University of Wisconsin, Milwaukee, WI 53201. The letter is dated October 9, 1979 and written on letterhead of The Association of American Geographers. I quote you the letter in full.

Dear Dr. Bowker:

We have now heard from Dr. Symanski regarding permission to reprint a significant portion of his article, "Prostitution in Nevada." Permission to reprint 18 pages and 2 figures from this article will be granted upon receipt of your check in the amount of $230.

The credit line should read, reproduced by permission from the AN-NALS of the Association of American Geographers, Volume 64, 1974, pp. 357–377, R. Symanski.

Sincerely,

Patricia J. McWethy

Executive Director

As you can see, Carville, the above letter is dated some two-and-a-half years *prior* to McWethy's letter of March 15, 1982 to Butterworths in which she censors my book and thereby opens the way for the leadership of the AAG to greatly expand the seriousness of the censorship.

You are also aware, your memory not failing, that the above incriminating letter was not part of the narrative in your hands, the very narrative which I invited you to control, manipulate and design within reason.

So this question: When "An Anatomy of Censorship" finally finds an outlet, do you think that even those people with only two-digit I.Q.s will fail to see that McWethy's salary, the cost of publishing the *Annals*, the cost of Executive Council members per diem (including, most likely, martinis and academic Gatorade) when they meet to engage in censorship, were paid, in part, by my time and my money—time and money spent to do the research for "Prostitution in Nevada," an article that appears, with few changes, as a chapter in *The Immoral Landscape*?

Here's hoping you enjoy your turkey and cranberries on Thanksgiving.

Part Five ◎ Inside and Outside and Eternal Postmodern Truths

Men seldom, or rather never for a length of time, and deliberately, rebel against anything that does not deserve rebelling against.

—Thomas Carlyle

Korski at thirty-nine

White Socks

> Nothing, indeed, but the possession of some power can with any
> certainty discover what at the bottom is the true character of any man.
>
> —Edmund Burke

WE called him White Socks because he always wore a suit and tie and white gym socks. He had a long and cavernous office on the first floor unlike any of the other offices anywhere in the building. When you entered and he asked you to take a chair in front of his desk, you immediately had a feeling of claustrophobia, of being too close. Perhaps it was the small space between his desk and the door, or his crowing, cackling voice, or the way his pockmarked face seemed to jump at you when he spoke. Or maybe it wasn't any of these things; maybe it was the incredible wall of books behind him that seemed to be tipping in your direction, about to bury you at any minute. So far as I remember, it was one huge bookcase. The few times I had a chance to run my eyes over the bookcase, I got a sense of having walked into an intellectual junkyard. White Socks had romance novels and cookbooks and academic monographs side by side, and he had old, yellowing newspapers under and on top of scientific reprints or folders that I'd seen him take to faculty meetings. What gave a special aura of incomprehensibility to his workaday environment, however, was a small sink at the far-south end of his office. In the poor light—White Socks had adopted that peculiar English habit of rarely turning on his overhead light—you could see that it had a huge goosenecked spigot and white tile faucet handles. On those occasions when I went to his office, there was a pronounced drip from the spigot that made a frightening noise. Once, while trying to explain a new course I wanted to teach, I brought the irritating drip to his attention. I said that it bothered me.

Forget it, he said as he dragged deeply on a cigarette, and put on a disarming comic mask that made him eminently likable in spite of some rather loathsome traits.

Doesn't it affect your concentration? I said.

On the contrary, he said. It's like music. It helps me write.

White Socks had once been the chairman of the department, but, by his own admission, he had given up the chair for "more important duties." By this allusion we came to understand that he hated all the paperwork, minor chores, and petty disputes that came with the job. We also understood that because White Socks had handpicked the new chairman, a person not known for backbone or decisiveness, he could continue to influence, if not in fact make, all the important decisions, the ones involving tenure, raises, hirings, and teaching loads. But it was always unclear to us younger faculty what White Socks had put in place of his administrative duties. He was now doing less research than previously, his teaching load had hardly changed (rarely more than one undemanding graduate course or two a semester), and his committee assignments in the department and the university were modest. Occasionally, he would ramble on about writing a book, but his talk was usually so vague and unconvincing that hardly anyone took him seriously.

But diminished output didn't lessen White Socks's own sense of self-importance, and he made it his business to know exactly what his younger colleagues were doing. He rarely asked us directly. Instead, he got all his information from the chairman, or the secretaries—always a good source of reliable information in university departments. White Socks's snoopiness and unwarranted arrogance were not at all appreciated by some of his colleagues. And on one occasion, his intrusiveness backfired in a way that I cannot forget.

Three years after I joined the faculty, the department hired a young hotshot who White Socks thought would make a significant difference in the ability of the department to attract good graduate students. I had voted against hiring Korski, believing that his personality was too strong, and that at the first opportunity he'd move on to a better university. I was certain that he had accepted our offer solely because he had not gotten offers elsewhere, a mystery, I must say, that I did not initially understand. At any rate, White Socks got his way, and I—at least—was glad that he had. Korski and I got along fabulously.

We shared adjoining offices, we frequently had coffee together, and when something went wrong that bothered one of us, we got in the tension-releasing habit of swearing at the top of our lungs in Spanish at the thin wall that separated us. (This example was about the only craziness of Korski that I adopted, and I did so because it made my home life so much more pleasant.) So, Korski's intellectual abilities and interests and ambitions aside, I was simply delighted to have him as a colleague. Too, my wife and I had just had our first child, and I desperately needed a summer salary. The university was cutting back on summer teaching, the market for my talents in night school had dried up, and I was having no luck at all getting outside funding for my research. Korski, wildly aggressive and always hopping about with a new idea, suggested that the two of us apply for a university summer grant for work in Bolivia. For the first time since I'd come to the university, the upper administration was offering "seed money" to anyone with an outstanding proposal. Korski said he had a surefire idea that could get funded; did I want to join him on the proposal? He would be the brains and write the proposal, and I was to do some bibliographic work, take care of fieldwork arrangements, and write any computer programs we might need to analyze our data. I was getting off easy: Korski would even write the first draft of papers that came out of our research.

Just before the proposal was to be submitted, the chairman called Korski and me into his office and said that White Socks had looked over our proposal and strongly recommended that we withdraw it. White Socks, the chairman said, thought the proposal was decidedly second-rate, an embarrassment to the department and its faculty. Korski, still in his first semester with us, laughed in the chairman's face. And then he quite forthrightly told the chairman that White Socks was full of shit and didn't know Spanish from Swahili and deserved a reprimand for sticking his nose where it didn't belong. The chairman, forever grateful to White Socks because he had hired him and then gotten him promoted all the way to full professor on the thinnest of academic records, became noticeably rattled. Not knowing what to do, he told us he'd see if he could get White Socks to change his mind.

Korski, I think, wanted to handle the matter himself. He didn't trust the chairman to convey his ire, and, as I would soon learn about Korski, he was never afraid of a confrontation when he thought he'd gotten a raw deal. But Korski held himself in check, though he did demand that the chairman send the proposal to the college irrespective of what White Socks said this time around. The chairman, I was not surprised to learn (nor was Korski surprised), did not send the proposal. Rather, the day after our meeting with him, he called Korski and me and White Socks into his office in an attempt to bring about a peaceful reconciliation.

I'll confess, I was scared. I wished I was sick or out of town, or could have found an excuse for not attending the impromptu meeting. Already I'd come to see Korski as volatile, and I'd heard from some of the graduate students that he was confronting White Socks in a seminar they were jointly teaching. This allegation perhaps requires a word of clarification. White Socks wanted Korski to run the seminar (for which White Socks was getting principal teaching credit) for two reasons: he—White Socks—wouldn't have to do any work, and he admired Korski's intelligence, thinking, quite erroneously it would soon become apparent, that Korski saw the history of the discipline and certain theoretical issues just as he did.

This assumption, of course, was a measure of how much White Socks thought of his own abilities. What he underestimated—rather badly, let it be said—was Korski's iconoclasm, his sense of unfettered intellectual combat, his much greater philosophical sophistication, and his superior grasp of the literature of the field. For his part, White Socks had truly blundered; he had not kept current, thinking, I guess, that what he knew twenty years ago and what he had gained through simply growing old would make everyone sit up and listen.

Sadly, even I could see that White Socks was a decade or more out-of-date. Other than being able to invoke some of the new jargon and drop a few of the right names, White Socks could not carry a first-rate discussion about philosophy or disciplinary changes with a good graduate student—or, alas, even with me. My world then and now was rather small and narrowly focused, and neither philosophical nor particularly theoretical. So it was hardly surprising that White Socks and Korski began to have disagreements on a whole range of disciplinary issues. And once you got to know Korski, even a little, you could bet that these disagreements would be aired in the seminar in front of the graduate students. White Socks, from all that I heard, did not fare well in these boisterous exchanges.

Well, to my pleasant surprise, when the meeting arranged by the chairman opened, Korski was calm and polite with White Socks. He asked him to explain his objections to the proposal and why he felt that it would embarrass the department.

White Socks couldn't come up with anything substantial, and at one point he pleaded that he couldn't remember exactly what had bothered him. I just don't think you've been here at the university long enough to be asking for summer salaries for both of you, White Socks finally said.

I didn't know that age had anything to do with good ideas, Korski said, still rather calm at this stage.

You think you've got a good idea, looking into the feasibility of crop-substitution programs for cocaine in Bolivia? White Socks asked.

Sure as hell do, Korski said, looking directly into White Socks's dark eyes. He pointed to me and said, He wrote his dissertation on Bolivia and knows the coca-growing areas as well as anyone. After finishing my dissertation, I spent several months in Bolivia following traders dealing in coca leaf. Furthermore, agricultural and drug markets in Latin America are interests of mine.

You're not going to pull back that proposal then, I gather? White Socks asked.

Never gave it a thought, Korski said. And then he added, leaning across the small coffee table and sticking his face in White Socks's lap, What gives you the right to be prejudging this proposal when you've never done a speck of work in Latin America? Far as I know, you've never spent a day south of the border.

White Socks fell back in his chair and pulled out a cigarette and smoked it about as fast as one can be smoked. Then, obviously angry, he got up and left without saying a word to anyone. That was the end of the meeting.

The chairman was utterly beside himself. As I recall, he actually wiped sweat off his brow (it was winter) and shook his head a dozen times or so.

Now I was scared. I thought, This looked like such a great deal, and now I might be losing my job. I wanted nothing more than to distance myself from Korski and find a way to tell White Socks that I had nothing to do with it, and if he thought it'd be better for the department to bury the proposal, then that'd be just fine with me. Unlike Korski, I didn't think that moving to another university would be easy—not at all. My wife was talking seriously about having another child, we needed a new car, and the last thing I wanted to face was the prospect of looking for a whole new career.

Anyway, Korski later told me that White Socks had pursued the matter with him in the hall that afternoon. He asked Korski to come to his office straightaway. Korski obliged, and White Socks used the occasion to verbally berate Korski and give him a dozen reasons why he needed psychiatric help. Korski, I would later learn, laughed in his face, and he told White Socks that in his opinion, psychiatrists don't have the foggiest idea what they're doing. He said, People who go to shrinks either have a two-digit IQ or have seen too many sophomoric Woody Allen movies.

Our meeting with White Socks occurred on a Tuesday, and though White Socks had classes to teach the rest of that week, he didn't show up for a single one of them. Nor did he show his face around the department. The chairman reluctantly submitted the proposal (I don't know whether he saw reason or was now as afraid of Korski as he was of White Socks), and we were funded. This resolution I was most thankful for, given my mortgage payments and the general financial condition I was in. I never heard another word about the proposal, or our successes, from either the chairman or White Socks.

The week after the proposal was submitted, I had lunch with White Socks in the cafeteria. He seemed unusually dour and pallid. It turned out that it had nothing to do with Korski. He had gone to the doctor complaining of chest pains. The doctor told him that unless he stopped smoking three packs a day, he had better start making out his will. The doctor had also told him that it might already be too late; he'd know for certain after looking at the chest X rays.

The medical news turned out to be good, and shortly thereafter White Socks returned to his normal self: lobbying for salaries based on seniority, control of all major committees by senior faculty, lighter teaching loads for senior faculty, and so on. He was not the chairman, but he acted as if he was—the same old White Socks.

Korski, meanwhile, began challenging White Socks openly: fighting for equality for all faculty, and arguing for a larger role for graduate students in departmental decisions. In one faculty meeting late in Korski's first year, White Socks tried to dismiss one of his arguments by saying that his attitude was not going to be good for his future. The threat was clear: Korski would soon be looking for another job if he didn't hold his tongue. Korski responded with a cutting remark, and then did what none of us had ever dreamed of doing. He threw a leg up on the table and pulled on a sock and said that he had asked Santa Claus for a pair of white socks. Just like yours, White Socks, he said. Broken elastic and all, he then added to rub it in. Korski, I regret to say, laughed and put on a supercilious smile while the rest of us sat there blushing with our faces in our hands. White Socks pointed a lit cigarette at Korski and shook it every which way, ashes flying. Then he jumped up and stormed out of the room. And that time, I swear, was the last I ever saw White Socks live up to his name.

I must confess that it was quite a shock to me, and so taken aback was I that I cut short my normally long Friday afternoon and went straight home to my wife and told her what had happened. She said she was glad that I had maintained my composure when we had had that fateful meeting with White Socks. She was even prouder of me for having had the good sense not to make fun of the way White Socks dressed. She was very reassuring. I could not, of course, tell her—and have not to this very day told her—that it was I who had first thought of calling him White Socks.

Korski stayed one more year and then left to take a job at another university. He left the department voluntarily, for a better job. At the time Korski submitted his resignation, I learned that White Socks had been busily lobbying the president of the university to have him fired. "Conduct unbecoming a professional scholar" was the charge.

Not long after the departure of Korski, White Socks came screaming out of the closet and, among other things, disclosed that he had been undergoing psychiatric help for years. He is now an avid promoter of gay causes, and, hard as I sometimes find it to believe, a staunch supporter of everything Korski represents. Indeed, these days he has nothing but praise for that maverick who had given him such a bad time. The only plausible interpretation I can give to this dramatic about-face by White Socks has to do with the sense of oppression and injustice that he felt in the university and beyond after coming out.

Pink Day-Glo Sneakers

There are braying men in the world as well as braying asses; for what
is loud and senseless talking other than a way of braying.

—L/Estrange

IN the fall of 1987, Nancy Burley, my wife, wrote to me in central Australia
and said that Kingsley Haynes was eager for me to apply for a position in the
geography department at Boston University. The previous year he had become chairman, and now he was anxious to build a first-rate department.
Nancy wrote, You should apply. King is anxious to get you. He thinks you
would make a big difference. He also said he'll actively pursue a position for
me in biology.

I wrote back to Nancy and said, Forget it. He's not serious; he only
wants to use me on one of his nineteen political agendas.

Apply! she responded. Despite what you may think, King sounds genuine. It just might work.

I'll think about it, I said.

21 March 1988

Professor John Fraser Hart
Department of Geography
University of Minnesota
Minneapolis, Minnesota 55455

(Version I: Not Sent)

Dear Fraser:

Since you were once again generous enough to write a letter of recommendation on my behalf (no doubt as strong and flattering as ever), I thought you might like to hear a bit about the recent job interview at BU.

It was an exciting and interesting trip, stimulating in a variety of ways too fascinating to adequately describe. I did not get to spend as much time with graduate students as I would have liked. But then, who can complain? They are, no doubt, as busy as bees in mating season.

My seminar on land abuse in central Australia was lively and provoked considerable interest. The discussion period lasted as long as my presentation (just about fifty minutes). I suppose a lengthy discussion is always a good sign, is it not?

I also gave a lecture on Aboriginal land rights to a large undergraduate class in political geography. I guess it can be inferred that they liked (enjoyed?) what I had to say, as they gave me a standing ovation when I finished. I suspect they were taken by the slides, the exotic subject matter, the personal touch. . . .

King seems as busy as ever, and I did not get to talk much with him about the department or where it was going. I did, however, hear about his designs for an elaborate geographic information systems lab with lots of hardware and software. At dinner the night before my formal seminar presentation he spoke of being involved in a national competition with UC Santa Barbara and Syracuse for some large monies from the National Science Foundation. He was as optimistic as ever. Getting money has always been one of his strengths.

Nancy's lecture at Harvard on the mating behavior of zebra finches went well (no surprise here), as did her discussion with biologists at BU. Who isn't excited by what she has found out about these little critters who weigh less than a slice of lemon?

As you might expect, we enjoyed Boston and its fine food. Now we'll have to wait and see what comes of the interview. I suppose I am guardedly optimistic. After all, who gets an academic job is always problematic, is it not?

Well, thanks again for your hearty support. No doubt it made a difference in my getting an interview, and perhaps it will prove decisive in the department's difficult deliberations.

(Version II: Sent)

Dear Fraser,

Fucked again. But then by now I shouldn't be surprised, should
I? Prior to going to BU I spent a good two weeks putting together a
faculty/graduate student seminar on land abuse in central Australia. The thesis of the seminar was that too many introduced animals, domestic and feral, are killing the land and life of central
Australia, and that although cattle numbers should be greatly reduced (even eliminated), the feral horses have to be shot in mind-
boggling numbers as soon as possible because they virtually alone
are destroying the remaining habitats to which native flora and
fauna have fled. Estimates on the percentage of native species that
have gone to extinction since the introduction of non-natives in
the nineteenth century run as high as eighty percent for small native mammals. Since abattoir demand for feral horses is slight, the
horses must be shot. I, and virtually all scientists in central Australia, would prefer to see all 250,000 feral horses in the Northern
Territory eliminated.

Nancy, wanting to make certain that I give my best ever performance, had me give her the talk three times, and she persuaded
me to include quantitative data that I would have preferred to leave
out. In addition, just prior to the trip, I gave the talk to biologists in
her department (Ecology, Ethology, and Evolution), and to the
wildlife branch of the Illinois Natural History Survey (Ph.D.'s in
biology who do a lot of first-rate applied work in the state). Nancy,
as I've probably told you a dozen times, is one hard critic, a brilliant
mind, and there was no angle of the argument that she allowed me
to get sloppy about. Anyway, the presentation was very well received in her department and in the Natural History Survey. (Their
questions came down to issues like costs of shooting and head vs.
heart and lung shots.)

Because Nancy had just given her seminar at Harvard and then
met with a group of biologists at BU (which, incidentally, had not
been told, contra King, why Nancy was really there to see them), I
encouraged her to find a cozy bar or restaurant near campus and
enjoy a glass of wine rather than sit through my presentation one
more time. But Nancy wanted to hear the polished product. (And
rib me if I left out something crucial, since I would lecture with

few notes). After the fact, it would be Nancy's assessment (and mine) that my BU seminar was probably the best one I'd ever given.

Before I'd gotten the last word out of my mouth, a tall, curly haired guy in the back of the room shot up his hand and aggressively and stridently let everyone know that: (a) I was completely wrong in my analysis of the problem; (b) I obviously understood nothing about the tragedy of the commons, and that this was a classic example of just this problem (Hardin's tragedy of the commons argument is, as you may recall, the linchpin of my last chapter in *Wild Horses and Sacred Cows*); (c) to shoot the horses or get rid of them by other means was clearly not the solution, and didn't I understand that ecosystems are "wholes" and to tamper with one variable like this was to disrupt the whole system; and (d) by eliminating horses it was certain that the ecosystem of central Australia would be pushed into a state of shivering disequilibrium. In short, where had my brains been crudely packaged, and where had I learned my sophomoric biology?

No mercy here, believe me. It was a good gut shot with a .22 short.

Well, I must confess that I was stunned by the aggressive attack. I played with my beard for a long ten seconds, then looked over at Nancy who was gritting her teeth and shaking her head and turning florid. For a brief moment I feared that she was about to spit out her retainers. She had a look of disbelief on her face that I've rarely seen in all the years we've been together.

But with more politeness than I thought I might muster under the best of circumstances, I went into some detail to try to convince my critic why he, in fact, had everything backwards and was sounding like a cranky animal liberationist who was irate over seeing his first white rat injected with a deadly human virus, an experiment that would eventually save his own son's life. I tried my best to convince him that this was clearly not a case of the tragedy of the commons. I attempted to explain to him just how it was that he was utterly mistaken to think that getting rid of a very destructive alien species was somehow going to push the system into a frenzied state of disequilibrium. I pointed out that I too had read Odum and Odum on ecosystems and knew all about Barry Commoner's First Principle of Ecology, but that here, among other things, we were dealing with the rapid loss of valuable gene pools.

The hungry young wolf kept shouting me down. Labels and gobbledegook and constant references to his mighty credentials (here he was less than two years out of grad school), not arguments or reason, were his ammunition. And how could it be otherwise? He had never set foot on the Australian continent. Nor had he ever done research remotely related to feral animals or the ecological effects of alien species.

Now Nancy was really starting to light up, and I was getting increasingly worried that this demure person with hardly an enemy in the world was going to turn around and spit one of her retainers at this bilious asshole three rows behind her. I was starting to think of the wine and seafood meals we'd have to forego if those retainers that were holding in her beautifully realigned teeth went flying and bounced off his quacky mouth and shattered on the floor.

Finally, when one of the wolf's own colleagues turned and informed him that he agreed with me that his understanding of the tragedy of the commons was more than a little confused, and then Nancy for the first time ever in an intellectual fray jumped to my defense, the ding-dong informed everyone that he had been an ecologist since the age of ten! And this, of course, was the reason that he was right and what I'd had to say was more than a little stupid.

Along about here my mind took off and magically flew back many years to a small village in southern Colombia (Tres Esquinas) and those unforgettable times when I was often in bed at midday with the most beautiful Indian girl I'd ever seen in my life, a girl who utterly captivated me, and there I was against all reason falling madly in love with this princess some ten years younger than me who in spite of little schooling had a mind as sharp and luminous as anyone I'd ever known, a mind matched only by her arresting high cheekbones, her flawless white teeth, a full mouth that defied description, and the loveliest glowing bronze skin I'd ever seen, and whenever we got it on she insisted on wearing nothing but pink Day-Glo sneakers that she had gotten from a nutty Peace Corps volunteer, shoes that so unhinged my sense of reality that because of her I developed this tic when making love of looking over my shoulder in search of pink Day-Glo sneakers bouncing up and down, a sight so unrelievedly comical that to stave off laughing and keep my mind on the delectable pleasures of the moment I'd try hard to think about why Mickey Mantle was the greatest center fielder of all time. . . .

Now you might think that this is the end of the story, but there was still more to come. For after the seminar was over (and the parry and thrust period was as long as my presentation) Nancy and I and King walked back to his office. On the way Nancy turned to King and said, King, I think you had better clip that first-year grad student's wings before he gets really dangerous and becomes a terrible embarrassment to the department. King, never one at a loss for words, nearly swallowed the knot on his tie, and even mine seemed to leap toward some unnatural part of my anatomy because I could not recall that Nancy had ever said something like this to anyone. Believe me, anyone. Nancy was, to put it mildly, absolutely pissed off. And I hadn't yet told her who the upstart with the fat mouth in the back of the room was.

Well, soon we found ourselves in what's known as the Pub at BU. There, some grad students and faculty and King and Nancy and I quaffed down several beers, and in the process I was reminded of that fervid religious element that one so often finds in American universities. In the previous two days the department had given me all of ten minutes to spend with grad students. But now, as we drank dark beer in semi-darkness, one young eager grad student with granny glasses turned to me and said that he had only one "serious and disturbing" question about all of my research, and that was why in *The Immoral Landscape* I had given a "sex" explanation to the demand side of prostitution.

He had trouble articulating what he meant, so I went to great length to explain why I saw the supply side of prostitution in terms of the economic oppression of women, and why I gave a biological evolutionary cast to explaining why men go to prostitutes. When I got through with all this, the student wanted to know why I'd missed the boat and hadn't explained prostitution in terms of gender, à la Friedrich Engels. I said that I had no idea what a gender explanation of prostitution à la Engels might be, and could he explain?

Well, unfortunately, the student was suddenly at a loss to explain either a gender interpretation of prostitution or where Engels fit into all of this. He wasn't the least impressed that my data base had been 300 "primitive" societies, data not available to Marx's old sidekick.

As the silence grew longer and I waited for this radical young upstart to powder me with Engels on the means and modes of production and class warfare and why I was really too bourgeois to be taken seriously, I began to daydream and, I hate to confess, my thoughts once again returned to that long ago time and I began wondering if God's finest gift to Colombia was indeed beautiful Indian women who could really fuck and always went to bed wearing nothing but orange or green or purple or pink Day-Glo sneakers.

Gonzo, man. I was really starting to trip.

Well, King and I—he had told me before my lecture—were to have one of those serious sit-down talks about, This is what we'll be able to offer.

But now there was no job talk that King was interested in; the verdict was in.

And so without apology he hurried off, to return to his sunless suburban castle and his new high-heeled coiffured wife.

Before he had reached the door Nancy snuggled up to me and said, Let's get out of here, I've had enough.

So we hurried out into the biting cold and hailed a taxi and headed for the wharf and some black pasta and crab-filled scrod and an expensive bottle of wine. Nancy and I now had not the slightest intention at this point—my stay at BU was finished—of talking BU or seminar or anything similar. I had no interest in wondering why the faculty there was so interested in why the discipline had "blackballed me" (their word, not mine), but didn't ask questions about my research. I had no interest in wondering why faculty at BU were upset with my "intellectual promiscuity." (What are you going to write on next? one asked. Aren't you going to just stay on the straight and narrow like good academics are supposed to? another asked. Etc., etc.)

But then somewhere between delicious mouthfuls of squid ink pasta, I told Nancy that the grey-headed wolf who had been a brilliant ecologist since the age of ten was the geography department's hotshot ecologist or environmentalist or whatever they're called these days. That he was a recent faculty addition who had gotten an offer at Clark and thought he might get one at UCLA and . . . well, probably would have gotten one from every geography department in the country if BU hadn't been so attractive and he couldn't wait for all the offers to come in that he just knew he'd get. I'd gotten this information from John Jakle a week earlier. John said that the wolf who wanted to chew me alive, Cutler Cleveland, was the best graduate student the University of Illinois geography department had "ever produced." Cutler Cleveland was, Jakle informed me, so good in the eyes of Jakle's colleagues that it was seriously doubted that the University of Illinois geography department would ever again have a student as good.

After I laid these tidbits in Nancy's lap she filled my wine glass and beckoned for the waiter and ordered another bottle; and then she gave me this one-of-a-kind unreal look, the identical look she gave me one snow-filled night in late January of 1971 in Syracuse after we'd been to one of those premier parties in the geography department where nothing happened. Anyway, now after an eerie, penetrating look came the words, You know, Rich, the only think I don't understand about you is why you want anything to do with geographers.

Tu-tu's Tale

> Burlesque and caricature best permit insights into Pueblo modes of
> conception since they reveal what the Pueblos find serious or
> absurd, baffling or wrong, fearful or comical about life and about
> other people.
>
> —A. Ortiz, ed., *New Perspectives on the Pueblo*

AT the Fourteenth World Wide Ecumenical Council held in Selma, Alabama (which included representatives from such diverse denominations as the Free Will Baptists, the Greek Orthodox Church in Canada, Sikhs of Syracuse, and the Open Bible Standard Churches), there was a small delegation from the International Federation of Marxist and Positivist Geographers (IFMPG). The delegation was headed by one Professor Swillock of doubly constrained entropy-modeling fame and soon to become chair of geography at Star of the Queen University, and included the distinguished Professor Brawling, representing Puerto Ricans, minorities, and the lumpen proletariat; Professor Gullgoer, speaking for anarchists, bald-headed Zen Buddhists, and small-business entrepreneurs; and the group's reigning shaggy-bearded guru, Professor Oxbalt, representing—so said the red-and-black button on his lapel—Karl Marx and Vladimir Ilich Ulyanov. Professor Rubicund, the tortilla-bellied squash player of the Santa Inez Athletic Club and noted student of unfindable maps in the head, was the most distinguished of the logical positivists present. It is rumored that at one point during the conference, Professor Rubicund was seen sitting alongside another famous positivist ("naïve and utterly paleolithic," Swillock would later say): Professor Aspen of Old Northwest University and formerly of the University of Great State, an expatriate who would one day become famous for calling friends and nonfriends alike in the middle of the night to ask for nonrepayable loans. According to the reporter for *USA*

224

Today, Dimitri Ahasuerus Bobaluka, noted gadfly and champion of radical feminists and anyone promoting queer studies, Professor Aspen was representing neopositivist wombats from the underworld.

Mr. Bobaluka (one day he would be Professor Bobaluka), an avid note taker of impeccable credentials and infallible principle, reported on a brief conversation among the IFMPG delegates that took place while Pope John XXIII smiled sweetly, recognized several new cardinals, played proxy for the queen in conferring knighthoods, and spoke of new calendars to help women recognize when they were "vulnerable." The pope also said a few words in favor of ecumenical embraces and hand clapping. According to Mr. Bobaluka's notes, the celebrated geographers present had the following memorable conversation.

PROFESSOR RUBICUND: My well-known position accords well with that of Paul Dirac, the world's greatest modern theoretical physicist after Einstein and Bohr.

PROFESSOR ASPEN: You would even put Dirac ahead of Heisenberg?

PROFESSOR RUBICUND: Well, certainly in that category. I suppose if you wanted to get contentious, we could throw in all the founders of the quantum theory: Born, Bohr. . . .

PROFESSOR SWILLOCK: Don't forget Schrödinger.

PROFESSOR RUBICUND: Is that where you got those fancy equations?

PROFESSOR OXBALT: You've forgotten Wolfgang Pauli.

PROFESSOR GULLGOER: That's right! Another brilliant German, surely someone we would have wanted in our esteemed department at Smallville.

Suddenly, there was a profound silence in the vast meeting hall. Sir Bertrand Russell, hunched over in flowing cape and walking with a staff, came slowly onto the stage. He briefly acknowledged the applause that was now deafening. He knelt before the pope, kissed his ring, and before anyone could gather their senses, he was made a cardinal of the church.

PROFESSOR ASPEN: Wow!

PROFESSOR GULLGOER: Who's that?

PROFESSOR ASPEN: You don't know?

PROFESSOR GULLGOER: Has he ever made a guest appearance on Sesame Street?

Professor Brawling, one hand firmly glued to his wampum-bead belt, shouted that someone should drop the damn screen so he could watch Young Frankenstein recolonize the colonized world.

PROFESSOR ASPEN (APROPOS NOTHING): The cape and the staff belong to Margaret Mead.

PROFESSOR GULLGOER: How can you tell?

PROFESSOR ASPEN: I just know and you should know. Really! By the way, have you ever read *Principia*?

PROFESSOR GULLGOER: What's that?

PROFESSOR OXBALT: The logical foundations of Marxism.

PROFESSOR GULLGOER: Where can I get a copy?

PROFESSOR ASPEN: Now, now! The blasphemy, the things you say at times! But . . . back to . . . what is the point that I've forgotten?

PROFESSOR RUBICUND: One you might want to remember in your future behavioral work. It is more important to have beauty in one's equations than to have them fit data. Those are Dirac's words. His exact words.

PROFESSOR ASPEN: You mean I've been getting too close to reality?

PROFESSOR RUBICUND: Sometimes you get so close I get edgy and uncomfortable, and then I can't hit the squash ball. Have you not learned that data will get you nowhere, only make you look foolish and simpleminded?

Professor Swillock smiled; Professor Oxbalt's face became incandescent—a deep burning scarlet; and Professor Brawling snorted and guffawed and said under his breath, *Viva la Revolución!* Professor Gullgoer, reigning king of Marxist thought at mighty Smallville, self-anointed anarchist and student of dialectical contradictions nonpareil, looked puzzled. Lost. Thoroughly lost.

Professor Aspen, ever eager not just to know the cutting edge but also to redefine its direction and velocity—a problem in vector algebra easily solved by someone with so much untapped talent—snuggled up to Professor Rubicund and giggled. And she said, Yes, yes! How foolish of me to be constantly seeking travel-behavior data from others to plug into my mathematical models.

Professor Brawling was temporarily distracted. Now he tried as he had never tried before to imagine what he would say when someone asked him what, as a good crypto Marxist-Stalinist, he was doing with the huge salary the state of Centralia was paying him to lead the university's Department of Geography in its quest for a thoroughgoing revision of its undergraduate program, one that would base all courses without exception on principles first enunciated by Marx, Engels, and, most of all, his real hero, Stalin.

Meanwhile, Professor Gullgoer scratched his suntanned, completely bald pate and wondered for the umpteenth time if he was on schedule for making Smallville the most famous geography department of all time. He was, he concluded, in a brilliant moment of insight, not just ahead of schedule but in fact there. The task had already been accomplished!

At precisely the moment Professor Gullgoer reached this insight, Professor Rubicund slapped his squash racket with the palm of his hand. He thought about swimming in the nude with one Professor Michael Sabbagh (of the University of Texas at Austin) and all those nubile camp followers always present, and where he was ever going to get enough money to own a kidney-shaped swimming pool with a slide and gold-embossed cocktail glasses from Kmart. He wanted one identical to that owned by the always generous Professor Kingsley Haynes and his lovely wife, Karen, when they lived in Austin-by-the-Lake and were king and queen at the KKH Ranch, which meant that one day both would be distinguished deans of American universities (not exactly first-rate universities, but universities nevertheless, which was all that mattered, all that ever matters to academia's technocrats and ladder climbers).

Professor Oxbalt, a genuine clairvoyant, annoyed at the crass, materialistic concerns of fellow geographers, had had enough. He reached into his two hundred-dollar leather shoulder purse, pulled out *Manifest der kommunistischen Partei*, scratched at his now graying, unkempt beard, and looked one more time for the key word that had been mistranslated and would, in the original, finally bring the teamster masses—just one of innumerable classes in American society about to take up arms and bring down the government—to their senses.

On seeing this display, Professor Swillock, who up to this moment had visions of negentropy dancing across his magical synapses, hit on a pipe and opened his sport jacket and out flew a *Pica pica* magpie with an iridescent blue-green tail.

Tu-tu, *Pica pica* among *Pica picas*, looked around at the possibilities and then hopped onto Professor Tumbleweed's lap, which was overflowing with absolutely first-rate *Annals* manuscripts. Tu-tu thought otherwise. He sat atop the spotless pile of wisdom that came express mail to Professor Tumbleweed from every corner of the world and relieved himself. It was a veritable explosion.

Attaboy, Tu-tu, Swillock said. Let him know what we think of his defection and his misguided predilections. Professor Swillock, let it be known for posterity, was thinking of the great promise that Professor Tumbleweed once had. Though he considered Professor Tumbleweed's mathematical diffusion models simplistic, not up to the mathematics that he had taught himself while sailing from London to Sydney as a young and aspiring geographer after getting a First at Cambridge, he nevertheless appreciated that Professor Tumbleweed had his mind where it belonged. Or once did, until that fateful day when he showed up at a renowned geegrogery conference and, to the chagrin and amazement of the conference's many Little Swillocks, discussed in captivating detail the quotidian mechanics of raising hogs in Oskaloosa, Iowa.

While all the above was happening, Lord Bertrand Russell came to the microphone and surveyed his audience. For several long moments, he ran a hand through his flowing white mane. He could not imagine what he might say to these estimable geographers. How great their minds! How unparalleled their concerns! What should he say? Twenty-four pounds overweight at ninety-seven, with black caves under his eyes, Sir Bertrand Russell looked exactly like Professor Kenneth Boulding—Mr. Systems Theory, Mr. Economics, Mr. Geography, Mr. Everything (or so it was widely alleged)—who was sitting between Professors Swillock and Oxbalt, the former petting the magpie and eyeing those *Annals* manuscripts, imagining a repeat performance.

When, finally, there was a measure of silence—only the yelping, irreverent Tu-tu could be heard in the whole of the great gathering hall—Lord Russell fixed his eyes on the International Federation of Marxist and Positivist Geographers and came to his senses. He realized that he had been given the distinguished, one-of-a-kind honor of standing before the most important single gathering of geegrogers ever assembled in the twentieth century. He had no choice but to talk about the only thing that mattered.

Lord Russell, let it be known for the record, talked persuasively about the need to blow up the Uranium Institute in Moscow. Then, wandering off onto themes that flashed to mind in no particular order, he arrived at an even more lofty plane (Dimitri Ahasuerus Bobaluka was madly noting all of it for posterity), something of more than temporal concern. He said, Mathematics is the only "science" (his hands were high in the air, his fingers twitching, to indicate quotation marks) where one never knows whether what one is talking about is true. Truth, it should be understood, having nothing to do with the imperatives of mathematical premises but rather with "empirical facts" (again his arms were high in the air), is that palpable reality beyond our skins.

Professor Tumbleweed, captivated by the speech, overwhelmed by the weight of the incontestable originality of the manuscripts that were making him feel light of mind, purred, I absolutely feel like a *Spelaeornis chocolantinus*!

While Professor Tumbleweed was contemplating his true nature, Professor Oxbalt reached into his bulging purse and filled his head with visions of Timothy O'Leary and jitterbug perfume. Then, making sure that no one was watching, he pulled out a copy of *Even Cowgirls Get the Blues* and began reading to himself with the utmost seriousness. He had found, he would realize many, many years later in the quiet of his leaky Cambridge apartment, that nothing in Marx could really compare with a little whiff of jitterbug perfume—or rather, better put, its image.

Several years after the Little Rock ecumenical conference, in a posh restaurant in Hamilton, Ontario, in the dead of winter, Professor Swillock, now easily the most influential geographer currently resident at Slavander University, justly renowned for all the carbon-copy Little Swillocks that he was so avidly reproducing, is sitting at a table across from a promising student who is thinking of doing her graduate work at Slavander. The student, obviously nervous, looks around the restaurant at the middle-aged other-kind-of-Indian women in furs and the stiff-backed men in Brooks Brothers suits. Her voice quivering, she says, You think we should have come here?

Professor Swillock reaches for the Molson bottle and fills his glass. He pets Tu-tu on the cheek with a bent finger and then says, What's your worry?

I looked at the menu. I'm not sure I can afford it, she says, saying all that needs to be said about poverty everywhere.

Professor Swillock takes the glass to his lips, gulps the beer, then puts his hand on the female billfold Tu-tu is standing on. Swillock fingers the thick stack of bills that he borrowed from his wife, herself a Marxist but one of questionable credentials, as she knows that Marx was a flaming chauvinist who bathed less frequently than Che Guevara.

I'll handle it, Swillock says. We get paid well in the frozen North.

No shit you do! Tu-tu screeches.

Swillock slaps Tu-tu on the rump and howls, Mind your bloody fucking manners! Then he says, Have you decided whether to come here?

The student looks at the white belly and the long beak of Tu-tu, then at a nearby fresco, *Sigismondo Malatesta,* by Piero della Francesca.

Tu-tu is motionless. He is fascinated by this student and her mysterious and unfathomable beauty.

The student glances at the fresco lovingly and wonders why she wants a degree in geography, and then, for reasons she will never be able to fathom for the rest of her life, she banishes the thought from her mind. Finally, she says, I've wanted very much to work with you. I'm confident that I'll be able to improve on some of the discipline's urban models. I think I've found a way to genuinely make them operational and falsify some of them.

Ooooh! Tu-tu crows.

Swillock furrows his grassy eyebrows, then screws up his mouth and reaches for a cigarette.

Tu-tu bangs his bill against the Molson bottle and then picks it up and drinks what's left.

Swillock says nothing to Tu-tu, or the student.

I really like the positivist work you and Professor Rubicund have been doing, the student says. But I've heard . . . I've heard you are interested in Marxism now.

Swillock takes a blunt thumbnail to his teeth and picks at the enamel on one. A tooth falls out. Tu-tu picks up the tooth and puts it in his hip pocket. Suddenly, Swillock reaches over and pulls out one of Tu-tu's tail feathers and then stands on the table and kicks him hard in the ass. Sitting back down, he says matter-of-factly to the student, Your source is pretty good.

The student says, Am I to conclude that you don't believe in what you've been doing all these years?

I might have gone down a different road.

You see all those years of work as leading nowhere, then?

A long silence, heavy breathing by Tu-tu (really, one of his many forms of laughter, if one understands Tu-tu's language and mannerisms).

Then Swillock says, his words measured, getting pissed off at the student's impudence, I wouldn't put it that way.

I really want to develop some good operational, testable, falsifiable models, the student says confidently. That's the problem I've had with the Marxist literature. With all due respect to what you're doing now, Sidney Hook and many others who have spent their lives studying Marx have shown that he is all wrong. Full of gibberish and inconsistencies, a historical curiosity. What's that word—*heuristic*? Marx, Hook and others maintain, is heuristic at best. Marx, Hook perceptively says, has little relevance to modern capitalist societies. Another thing, a couple of years ago I went to a conference in which some Yale scholars (I don't recall their names right off) combed the literature and found that very little in Marx has been tested—I mean, empirically tested. This divorce from reality doesn't really suit my personality.

Tu-tu, now sitting on Swillock's shoulder, cackles. Then, loud enough for people to drop their forks and swallow their baked salmon whole and choke to death, he yells, Right-ooo! Right-ooo!

You brash little fucker! Swillock yells at Tu-tu, bringing his foot up onto his shoulder and kicking the *Pica pica* magpie in exactly the same spot he'd kicked him before. Turning to the startled and frightened student, Professor Swillock now says, You believe you'll get closer to truth in the positivist tradition, then?

Their models and theories are testable. At least when they don't work, they can be rejected.

His face the color of smilax, Swillock laughs, reaches for the empty beer bottle, and drops his cigarette to the floor.

Tu-tu, who's had enough for the moment, hops off Swillock's shoulder and releases a blast of wind so mighty that the front window falls out of the four-star restaurant.

Like nothing has happened, Swillock says, I'm beyond *Troilus and Cressida*. Then, grabbing several pieces of calamari and stuffing them into his mouth, he says, You need to do some more reading. Start with Lakatos and Bhaskar.

Swillock eats what's left of the calamari, orders another round of beers, looks at the bill, and puts three twenties and a ten on the collection plate. As he does so, Tu-tu pulls out Professor Swillock's well-worn rosary beads. They fall to the floor and make a loud noise, and Tu-tu lets out a screeching howl, pure laughter in another key. The waiter comes over and tells the distinguished Professor Swillock, a frequent customer, that if he doesn't make the magpie behave, then the bird's going to be on the menu.

Ha, ha! Tu-tu cries. You'll get yours soon enough.

Several weeks later, the student, eager to engage the fast-emerging Marxist geographer, gets a wad of bills from her boyfriend, Korski, and puts on her one-and-only pair of pink Day-Glo sneakers and a Michael Jackson glove on her left hand and gets Professor Swillock to agree to meet at the same restaurant for Molsons and Black Sea caviar on rye.

So, now I see what you've decided, the student begins. You have shown that your work and Professor Rubicund's work and Professor Casetti's work and the work of legions of mathematical geographers—the list is awfully long, isn't it?—were all thoroughly protected by *ceteris paribus* clauses. Nothing could be rejected. And if a theory could be rejected, then no one wanted to.

Swillock nods.

Tu-tu nods.

You liked it that way?

Swillock is silent. Silently pissed off.

No sense of how things should be?

Now Swillock is more pissed off than ever.

Tu-tu laughs, picks up the Molson bottle, and has a hearty drink.

Its justification is that it was practice, right?

Swillock nods.

Tu-tu lifts his tail feathers and says what must be said in the only way it can be said.

So now you see what you then did as almost identical to what is going on in Marxist research? Masturbation with words and more words, instead of alphas and betas and gammas—if you don't mind the sexual metaphor—justifies all? No idealism about rigorous testing, that we ought to search for truth, what practice should be? Positivist geography didn't pay attention to facts; Marxism doesn't pay attention to facts. Big deal, you say? All that matters is theoretical twining.

Swillock shows no emotion. He's trying to imagine ways to strangle this upstart student, beautiful and charming as she is, he must admit to no one in these changing times. To calm his nerves, he orders four bottles of beer, more caviar, and more rye, and he says he wants to see a menu. Eat to calm, if not kill, the passions, he thinks.

Tu-tu picks up Swillock's Molson bottle, finishes it off, throws it at the waiter, and hits him in the head. The intentional hit is fatal. But Professor Swillock, with much more important issues on his mind, hardly notices.

If, as you claim, all facts are problematic, the student continues, then I don't understand this Lakatos thing, this thing you and your wife agree to—about the value of the untested theory predicting novel facts. You don't recognize facts, yet you're going to predict novel facts! So the only facts that are not problematic facts are novel facts—is that right?

Professor Swillock, every bit the scruffy, beer-guzzling Aussie at heart from all those memorable, swilling Canberra days, lets go with a mighty explosion, equal to any for which Tu-tu is justly renowned.

Without apparent provocation, or reason that is subject to reasonable analysis, the student from Tres Esquinas by that river with changing names unties her pink Day-Glo sneakers and takes them off, and they rise on their own and begin dancing as only sneakers can dance, directly above Professor Swillock's head.

The student, visibly disturbed, a victim of forces now utterly beyond her control, worried beyond words about her future in the greatest discipline of all time, meekly says, Sorry . . . sorry . . . and for putting it so bluntly. And . . .

Tu-tu cries, Ha, haaaa! Then he yanks off the student's silver-studded

Michael Jackson glove and eats it in one bite, thinking as he does so, Fucking heretic!

Swillock, knowing his bird as he knows nothing else, kisses Tu-tu on the cloaca and says, Good boy, Tu-tu. You are a bird of my heart.

Another thing seems to follow from all of it, the student brashly says. You turn to Marxism because, you say, there is this ineluctable connection between theory and praxis, but there's no praxis that I can see. Can there be praxis without getting out there? Without living differently? Excuse me for saying so, but . . .

I have puzzles to solve, Swillock says calmly. He sits back, stuffs his face with caviar, and, in a moment that would even shake unshakable Marxists, drinks two bottles at once.

Ooooooooo! Tu-tu croons.

The student, now fearing that she's gone way too far, yet wondering how she can get to the many other problematic arguments and blatant contradictions she sees in Marxist academic practice, leans back in her chair. But suddenly, her chair lifts off the floor and disappears, and she falls to the floor and notices that she's absolutely naked.

Ooooooooooooooooo! Tu-tu croons louder than ever, as he finds his master's ear and, following some primitive law of nature yet to be discovered, bites it off. The crooning shakes the diamond chandelier in the middle of the restaurant. Ooooooo—, and the chandelier explodes into a million pieces, and the five customers still in the restaurant die almost instantly from flying glass, and the shock of the unexpected.

Professor Swillock, bleeding profusely and looking for the ear that Tu-tu has eaten, has had enough. He stands and reaches back for all he's worth and takes a deadly swipe at Tu-tu. He barely misses Tu-tu's neck.

Oooh! Tu-tu mourns within himself. I have gone too far, he thinks. I'll never again be able to play with my master's rosary beads, or make fun of his upstart graduate students. Disconsolate, knowing that all is lost in his captive world, Tu-tu flies off. He flies through the ceiling and straight up into the stratosphere, and in mere minutes the long-tailed black-and-white member of the best-known species of the family Corvidae, of the order Passeriformes, who is usually content to eat seeds and the eggs of the young of other birds—but also loves carrion red and ripe—gathers all of his kin in North Africa and across the whole of Eurasia and in every nook and cranny in North America, and like the plague of the Bible, they fly en masse with the jet stream and bomb the restaurant with hundreds of millions of alien bird eggs. Professor Swillock and everyone in the restaurant—save the student, the lucky, lucky, incomparably beautiful student with the pink Day-Glo sneakers—are buried so deep in shells and the unborn young of birds that not the slightest sign of a human can be found in the aftermath.

And that event would be the end of the story, were it not for the fact that many hundreds of years later, the immortal Dimitri Ahasuerus Bobaluka, the most brilliant fence-type and house-type and cemetery chronicler ever produced by the state of Georgia and for decades the reigning, self-anointed king of *USA* map mapping at Snow Mountain State, would relate how that one evening, high in the brassy sky, he had seen a single *Pica pica* magpie with a pair of pink Day-Glo sneakers dangling from his beak. And this scene, he concluded, was a sure sign that some kind of truth ("some kind of truth" are the exact words found in his notebook) will most certainly prevail.

Korski on Symanski, Symanski on Korski

> What has brought about this change? One thing and one thing only:
> Where previously I strove to cling on to a host of things, now, when I
> have lost hold of them all one after another and have nothing left but
> myself, I have at last regained a firm footing. Under pressure from all
> sides, I remain upright because I cling to nothing and lean only on
> myself.
>
> —Jean-Jacques Rousseau, *Reveries of the Solitary Walker*

WHAT'S up?

You didn't finish the job.

I thought we said what we had to say.

I thought so too. But two of my faithful communicants, Quetzal and
John of Oz, have told me that if we don't explain where we're coming from,
we're going to be ripe fruit for easy picking. And since the fruit's got a pretty
strange taste, all kinds of spittle and misunderstanding are likely to come
our way.

Yeah, I hear you. But I thought you and I lived by that principle of good
fiction that says: Less is more. Don't give them everything. Let them do
some work on their own, read between the lines. It's not only more fun this
way, but it's also . . . shall we say, more pomo PC too. The more voices the
better.

Well, you'd be surprised what the masses can't see until you put it right
in front of their noses and spell it out.

Or don't want to see.

Whatever, I think we ought to take Quetzal and John of Oz seriously.
I've got a gut feeling they're closer to the ground and reality on this one
than we are.

Well, I suppose.

You know that doing this can work against us?
And for us.
Blind optimist!
Okay, okay, who's going to do it?
You.
No, *you.*

My career in academic geography effectively lasted about five years, from the fall of 1971 to the spring of 1976. I was a tenure-track assistant professor at the University of Cincinnati from 1971 to 1973, and the same (except for the first year when I had a visiting position) at the University of Texas at Austin from 1973 to 1977. Were it not for a year when Michael Webber (long the chair at Melbourne University) was visiting and in the department, a time when I thoroughly enjoyed numerous intellectual free-for-alls with him, I could claim that in terms of intellectual discourse, the Department of Geography at Cincinnati at that time was an intellectual wasteland.

There were telling incidents. One day early in my first year, I asked a colleague, Bruce Ryan (later chair of the department), if he'd share with me what he'd thus far written for a book he was allegedly putting together on Appalachia. At the time, I was preparing to teach a field course based in eastern Kentucky. Bruce reacted to my request as if I'd publicly questioned his manhood. He wanted to know where I'd acquired the exceedingly bad manners to ask such a question. Then, in his own Anglophile quippish way, he proceeded to tell me that whatever he was doing on Appalachia was none of my damn business. I never spoke to him again about anything remotely related to an intellectual issue.

Howard Stafford, the chair, once, and only once, asked me to critique a paper he had written and was about to submit for publication. I spent a good bit of time on it, and I gave him a couple of pages of written reactions, including suggestions on how to improve it. He was visibly upset with my critique, and it was clear that I had ignored his implicit instruction to me, which was: Just tell me how great it is! Stafford subsequently took me aside and said, "You want to get along, go along."

For challenging the authority of Bob McNee (the de facto head who loved having someone else do his dirty work) on intellectual issues, for not endearing myself to the chair, for taking issue with the university's elitist honors program while sitting on a university committee, and for dressing in a way that Bob McNee considered inappropriate (bell-bottoms and moccasins), I became persona non grata at Cincinnati.

My second and last tenure-track geography position, at the University of Texas at Austin, was even more dispiriting. With the exception of a small handful of economists, historians, and anthropologists in the Institute of Latin American Studies where I held a joint appointment with geography, intellectual stimulation was not easy to find. And the last place to look on the Austin campus in those years was in the geography department.

During my four-year tenure in the department, I can recall only a single substantive conversation that lasted more than five minutes, and it involved a young member of the geography faculty by the name of Ian Manners. This conversation occurred only because I needed to spoon-feed Manners on nomadism and periodic markets so that he could write up a section or two and be included on a paper conceived and largely written by me and R. J. Bromley. Manners badly needed the paper to add to a very thin publication record. Without it, promotion and tenure seemed chancy. He would later thank me for this needed *Annals* freebie (after he got promoted and tenured) with an act of monumental ingratitude.

I still vividly recall reprints from articles I'd just published appearing in my mailbox, or in the departmental office garbage can, a day or two after I'd given them to colleagues. My so-called colleagues were more than a little judgmental about my "hot" research topics. I was roundly taken to task for having the "audacity" to critique, at the University of Minnesota and in the *Annals,* that "very esteemed historical geographer" D. W. Meinig. I was upbraided even more so for having the "gall" to write about prostitutes in Nevada. There was not a smidgen of interest shown in the content of what I had found, or my field methods: everything was about my transgressions of unwritten tribal social norms that I should have known about.

Following the quick appearance of several articles in major journals, and invitations to give presentations at major universities, I asked to be put up for tenure a year early. I expected one negative vote because I had confronted one faculty member for being extremely rude to my girlfriend (later wife). I heard nothing until one Monday morning when another member of the faculty asked me to come to his office. When I entered, Paul English—like McNee, the de facto chair and departmental opinion maker, power broker, and rule enforcer about everything that mattered in the department—was sitting quietly next to the chair. Together (two rather than one equals more authority, a more definitive show of power) they informed me that the previous Saturday night, while gathered around Paul English's hilltop swimming pool, they'd gotten a seven-to-zero departmental vote to terminate my contract. Among twenty-nine charges leveled against me (several were more whole-cloth fabrications and more surreal than my wildest fictions) was the central charge that I wasn't a Latin Americanist, the position for which I'd been hired. My activities at the institute (then just about the best in the country) were irrelevant, as was the fact that in the short time I'd been at Texas, I already had more of my published articles reprinted in the institute's long-standing Latin American Reprint Series than anyone in the history of the department. My excellent teaching evaluations also counted for nothing. With this completely unexpected Saturday-night assassination, I had been reminded that hate, power, and envy in the academy have no limits.

During the fight that ensured over the vote to fire me, and fight the vote and the department I did, the dean of the school, Bob King, asked me point-blank (his exact words): "*Who* are those fucking aborigines upstairs who have treated you this way?" A college tenure and promotions committee in its only deliberation of my case quickly and unanimously overturned the vote of the department. Following this decision, the dean again spoke with me, this time making it quite clear that although I had won the battle and made the geography department look harebrained in the eyes of many in the university, the department had won the war. He told me that I was now like a whore in a convent, and I'd better look for an academic home in another university as soon as possible.

I never really found one. After I left Texas, I had a part-time, two-year visiting position in the geography department at McGill University. The one saving grace of my stint there was befriending and spending afternoons chatting with the activist and peripatetic arctic anthropologist Hugh Brody. Again, as with my previous academic appointments, I could usually find more stimulating conversations in blue-collar bars than I could in the geography department.

In my second year at McGill, I got another good taste of how power works in academic departments and universities, historically almost solely to the detriment of women up to that time. The chair tried to double my teaching load, and at the same time cut my salary by almost 20 percent. I refused to accede to both, and my guess is that had my wife and I not left McGill after that year, there would have been nothing more for me in the geography department.

My academic stock, my publications notwithstanding, was heading south faster than Canada geese in the fall. By now, I was repeatedly hearing that I had acquired a "bad-boy" and "troublemaker" reputation. I was sometimes paired with the infamous Bill Bunge, as people to avoid. It was all my own fault, I was being told: for taking D. W. Meinig to task in print, before I had tenure; for having the "stupidity" to write about Nevada's prostitutes in the same journal, before I had tenure; and for being even more brainless in fighting the Texas mob rather than just closing my mouth, bowing my head, packing my bags, and leaving town on the first bus out. I received this judgment from some of the biggest stars in the discipline.

I more or less followed my wife to the University of Illinois because she'd been offered a tenure-track position in biology, and because I'd been promised by the chair, Art Getis, that there'd be a position for me in the geography department. But Getis's promises were all smoke and mirrors, and reality quickly proved to be something I couldn't stomach. Getis "honored" me with a basement corner hole called an office that had neither bookcases nor even a wastebasket—and, of course, I received no salary. Furthermore, some faculty had the audacity to ask me to give free lectures and then supply them with exam questions on what I'd said.

Just about everyone in that Illinois department in those days had no interest in talking with me about the book I was finishing on prostitution, *The Immoral Landscape.* My colleagues were interested only in letting me know, principally behind my back and by way of gossip from unnamed sources, that I obviously had nothing new to say, because anything worth knowing about prostitutes had already been written. Worse, with this wayward if not perverse interest of mine, I was very bad press for the department and the university. The only person to my knowledge who would genuinely have no part in this uptight blather was John Jakle, a genuine and principled gentleman.

Before long, I was told quite explicitly that because my wife had a job in biology, I was trapped. I was ripe for being "exploited," and that's what the department had in mind for me. Angry beyond words, even regretting that I had left Texas rather than continue the war, I went home and looked for another life. I wrote, and for a brief time warred with the AAG over the censorship of *The Immoral Landscape,* a conflict that added only fuel to my reputation as an obstinate troublemaker. I read scores of novels. I worked on my writing skills. I developed an interest in photography. I traveled the West. I wrote three novels (all near misses with the publishers, which means they might as well have never been written). I wrote a book on wild horses.

Now well seeded with cynicism about geography and the whole academic enterprise, I began having fun. I wrote essays and screeds that mocked and ridiculed geographers and what they were doing: pieces with titles such as "Tu-tu's Tale," "White Socks," "Teeth," "David Harvey on the Lettuce Trail," "Looking for J. B. Jackson's Ordinary Landscapes," and many more. How many I wrote I don't know, as I didn't save copies of several of them. Some I wrote under the name of Richard Symanski, some under Korski. A few were unsigned; a couple were penned with names taken out of thin air. I never thought of publishing any of these pieces. It was satisfaction enough to have written and sent them to half a dozen or so people—a few friends, and geographers I wanted to provoke and irritate. I didn't expect feedback, and I didn't get any.

In January 1994, in the process of moving my tenured line from a social science unit to biological sciences at the University of California at Irvine, if only to "erase" my past while pursuing conservation interests, I began a roaring e-mail correspondence with four of the people I have identified in the text: Jacquot, John of Oz, Quetzal, and Don Pietro. Over the next several months, in fact until June 30, some two hundred thousand words passed among us, most of this output coming from me. The missives and messages were unpublished papers, sketches of trips I'd made to the Colorado River and to Mexico, fictions, missives sent to university administrators, essays on my growing son (then three going on four), and several pieces similar to the ones I wrote about geographers and the geographic enterprise in the 1980s. They were irreverent, mocking, analytical, and serious. The reactions from my friendly communicants were sarcastic, playful, insightful, and dead serious about too much that's wrong with universities. It was a marvelous and unprecedented time—and very satisfying. We had a lot of fun, we vented, we became better friends.

But then, feeling that maybe I ought to shift gears and do some serious conservation work, I woke one morning and sent out a note to my friends saying that Korski had died of a pistol shot, and that funeral services would be held for him at Charley Utter's Camp on Thursday afternoon, July 15, 1994, at three o'clock. "All are respectfully invited to attend," I ended.

Soon, however, my new life as a biologist just down the hall from my wife began to take unexpected turns. I'd ventured into a new nightmare. I was swiftly reminded that I didn't have the right union card—a Ph.D. in biology—and therefore couldn't be taken seriously. I was told that, contra everything I knew, conservation biology had nothing to do with human institutions and behavior. A colleague with a nearby office let me know that he'd be happy to see the Humanities Building on campus go up in flames.

And then began a battle that lasted nearly three years. Contra my tenure papers and contract with the university, and in the face of just about the most telling evidence possible—a single-authored, peer-reviewed book in biology at a major press *(Blackhearts: Ecology in Outback Australia* [Yale University Press, 2000])—I was repeatedly informed that no matter what I wrote, I wouldn't get an ounce of credit for it anywhere in the university. I fought the issue with two chairs, three deans, two executive vice chancellors, two chancellors, and the president of the entire University of California system. With several inches of argumentation and documentation supporting my case, I was my own attorney. But I was not good enough, for I lost, managing to do no better than get a formal written "rebuke" and be put on three years' probation for the aggressive and, yes, patently just war I'd waged.

I now concluded that it would be utterly masochistic to bring anything I wrote onto this campus where I taught. I now concluded that I'd get more mileage for anything I wrote by printing it out and selling it for worthless pesos to Cubans who had run out of their *Granma* toilet paper.

I took all of my research (a couple of articles in progress, a book half completed on salmon, and another one outlined and under way on agriculture and water in the Imperial Valley), boxed it, and pushed the boxes into dark corners in the garage. I returned to an old hobby, photography. I began spending all my free time in the university photo labs, and with art students. And I headed south, literally. I began driving to Tijuana on a regular basis, almost always to Zona Norte, the red-light district. I walked the streets day and night. I took thousands of color and black-and-white photos. I learned more about prostitution in Tijuana than I ever knew about it in Nevada or anywhere else. I didn't have an ounce of interest in writing up any of this decent ethnographic data.

Then one night, in the heart of the red-light district, I got mugged: choked into unconsciousness, dragged into a dark alley, and stripped of everything—cameras, credit cards and money, glasses, keys, belt, even loose cough drops in a shirt pocket—save my pants and shirt. I was left for dead not fifteen feet from hustling food vendors and a dozen sleepy-eyed street prostitutes who had idly stood by and watched (with indifference, if not amusement, I'm sure) the two thugs who took me down with a professional choke hold and then cleaned me out.

Less than a year before the mugging, I had rediscovered Cuba, a place I had first tried to get into while at the University of Cincinnati. Initially, I spent a month there to shoot a couple of thousand photos of people and street scenes. But from the first hours after setting foot in Havana on my first trip, I couldn't stop asking questions, hitting on every Cuban I could find to talk to me: about the economic deprivation, the parallel markets, stealing from the government, perceptions of Fidel, the suffocating political environment, the hard-hustling *jineteras*. Before long, and after a second trip to the island, I had developed an ornery kind of love affair with the neglected land and its high-spirited people. Now I wanted to write a book about Cuba. I figured that if I got it published, then great; if I didn't, then no big deal. After all, beyond teaching a few no-brainer courses, life now was all about my hobbies, my addictions, my need to write, my son, my wife.

I got five chapters into the book on the island prison, all in hand before my third trip in mid-1999. But I wasn't satisfied with where it was going, what I knew about certain issues. I wanted more data, I wanted a deeper understanding of the black market, the *jineteras*, the quotidian lives of very ordinary Cubans. I put these chapters aside, began planning for another trip to the island, and returned to printing my Cuba negatives, putting together a show.

I continued to write essays on anything that caught my attention—just to be writing, to satisfy this needy addiction that reveals the limits of my mind, my imagination. For reasons that I don't completely understand, the mugging experience in Tijuana in mid-January 1999 had suddenly put me into a feverish writing mood, unlike one that I had previously experienced. I was now on my second or third life, I told myself. In the months before and after a long trip to Chiapas, and Cuba again, I wrote a couple of short stories, and more than a dozen essays on my son: his experiences with baseball, learning to read, what I wanted to tell him about affirmative action, why it was the right thing to do to bomb the Serbs, how I now remembered my days of glory as a star basketball player. All of these essays were headed for my black file cabinets, for viewing by my son ten or twenty years thence.

Then one day, Jacquot, a friend who goes back thirty years, sent me an e-mail suggesting that I read an exchange in the *Annals* on Ed Soja's book *Thirdspace*. I read it, and it reminded me of what I'd more or less left behind: the bald-faced bluffing, the pompous posturing, the woolly and lousy writing, in general the maladies and the phony liberals of academia that I could now pretty much ignore in spite of getting my check from a university.

A long day and a morning of writing, and I had a Soja essay, a piece to send back to Jacquot and a few other friends, and to Soja and his friends. Then I followed it with a piece on Don Mitchell when Jacquot once again said I ought to read Mitchell's book on California agriculture. Mitchell, after all, was now a MacArthur Award "genius," and with my love and admiration for giants and geniuses, I just couldn't resist looking at what Mitchell was up to regarding bent-back fruit and vegetable pickers in my native state.

Then I got my copy of the *Geographical Review,* Starrs's loving homage (via carefully selected, same-minded, adoring acolytes) to J. B. Jackson. I carefully read the hagiographies, and I read the misguided piece by Peirce Lewis on how to do ordinary landscape analysis. It was time to return to what I'd written more than ten years earlier on Jackson and couldn't get published because it went too much against the received wisdom among cultural geographers about Jackson's singular greatness. It was time to return to my good friend Korski, my constant soul mate, now once again alive and well.

The sport now picked up speed. Willy-nilly, I'd think of a high-profile geographer, then e-mail the library to put one of his books in my university mailbox. I'd bring it home, read it, then spend a morning writing another essay, if I wasn't teaching writing that morning to biology majors. I was on a roll: Dick Walker on Don Mitchell, Curry on words in the world, Jim Duncan in his Kandyan confessional, Bob Sack and *Homo Geographicus,* John Fraser Hart on rural landscapes, Saint Lefebvre and *The Production of Space,* Richard Rorty on campfire pragmatism, Peirce Lewis on New Orleans, hot-wind Paul Starrs on overgrazing, and more. I sent these essays to my faithful communicants, and to a few others, and always, and right away, a copy to the person I had written about.

A few people e-mailed me and said, Don't send anymore; they're too strong—I can't handle them. A few said, Send the one about so-and-so that I've heard about but haven't seen. A few wrote and said they sure hoped they weren't on the list of geographers I was lining up to whack. My communicants said, Jesus, Korski, what next? Or, Great, about time! Or, When are you are going to go after Pred—he ought to be an easy one? John of Oz always had a lot more to say, adding commentary, his sense of context, pointing out the missing apostrophes that I was supposed to be showing my students how to use. Don Pietro, even as he lay dying, sent me line-by-line corrections, suggestions on where I might elaborate to refine a point. And he sent me a copy of *The Big Book of Filth* to enrich my impoverished street vocabulary, much as John of Oz about the same time airmailed me a book called *The F Word.* Dick Peet told friends and comrades to ignore the "madman" Symanski. One tenured geographer in a prominent department whom I took to task in one of the essays in this book threatened to smash me in the face the next time he saw me; he said he was a professional boxer. He was someone I had given the benefit of the doubt to, treated kindly.

But the whole experience of this on-again, off-again hobby, writing these adrenalin-rich essays, usually in one sitting, was getting depressing. And a bit boring. They all seemed too easy. I began rhetorically to ask myself why there was so much dumb, idiotic, indefensible, or pompous twaddle, why no one else was exposing all these pretenders, frauds, and second-rate thinkers. For the umpteenth time, without showing her what I was writing—only now and again telling her more or less what I had written—I'd ask my wife what she ventured was wrong with this discipline called geography.

I'd get several familiar answers, and then her stock answer to a follow-up question about me. You're third standard deviation, she'd say—just as she often says when referring to what I write, my meandering travels, the risks I take, my oddball interests, the reach of my mouth. She means, by the conservative straitjacket norms of academia and its long reach, you're—yes, Rich or Korski—mad, a deviant, utterly hopeless. It is the friendly and loving version of what Dick Peet and others have said, I suppose, not knowing more than a fraction of what is said behind my back by people without even enough courage to sign their reviews of submitted journal articles.

So you might finally get fifteen minutes in the sun with this one?

Won't be any longer, I'd bet.

You mean you don't expect any dinner-party invitations from Ed and Jim and Paul and Peirce?

Well, as a matter of fact, I'm looking forward to several. And if not that, then I at least hope to be able to get wind of their spicy descriptions of what they think of me as a person.

Oh, they'd never speak poorly of you. Their responses will be purely intellectual, objective. Getting personal, along with using a choice four-letter word or two on the printed page as you're all too willing to do, is one of the great remaining taboos in the academy.

Yeah, but they'd just be returning the favor. I mean, I've been nothing if not personal, as you will soon hear when you read the reviews of this book. You see, one of the things I forgot to mention about the Meinig brouhaha is that the most consistently damning charge against me in the '70s was that I had been personal. That if only I had also, say, written about how David Lowenthal and Yi-Fu Tuan manipulated language, and done so in that same article, then no one would have said boo about what I had said about Meinig.

No, you're putting me on! *Everything* is personal when you don't like what someone has said about your work. And, anyway, what about all this stuff about all reality being socially conditioned? You mean, the social is not the personal? You mean, we can talk about how everyone is socially conditioned for what they write and represent, but we can't talk about how a person's love of English canes or cowgirls or distrust of authority or fascination with prostitutes affects what he writes? I don't get it. . . .

I don't either. Never have.

Write an essay on it.

Got better things to do.

What about others, the ones you haven't dissected and diced? How do you think they're going to respond to these essays, embracing the methods of combative journalism, making fun of people?

Variously, I'd say. Just about everybody likes to see a little blood flow. Did I ever tell one of the great truths that John Fraser Hart passed on to me while he was editor of the *Annals*? He said, You know what people turn to first when they get a new issue? The commentaries. Of course, none of it can be admitted publicly. No one wants to admit that he's at least one part savage, or intellectual sadist, and maybe even a happy cannibal if hungry enough and no one's watching. Chagnon's not all wrong about the Yanomami, believe me. I'd guess there'll be plenty of geographers who will be delighted with what I've said about Ed or Paul or Dick or Don, and there may even be a certain amount of private gloating about someone getting his comeuppance. But tell Dick or Don or Peirce how they feel about what I said here or there to anyone other than the mute mirror, the faithful wife, a very trusted friend? You gotta be kidding. Academic geography is more incestuous than a widowed father with a gorgeous daughter on Pitcairn Island.

Sounds like you might get more than fifteen minutes after all, you unrepentant savage!

Now that I think about it, maybe twenty, a little more. I know a few guys out there who'll give this book to graduate students and say, Look at this, then come and tell me what you think.

Very objective, neutral-like, right?

Yeah, just like my eyes when I watch a beauty pageant.

You might even make it into a graduate seminar! Come to think of it, I wouldn't be surprised if some bright-eyed, go-for-it graduate student tries to whack you a good one for what you've written.

Wouldn't be any game on this end if someone didn't try to do a number on me. At least someone would be taking me seriously—for a week or two.

What do you mean—a week or two?

All right, three weeks, a month. Then the student will come to his senses. He'll think: Well, that was good fun for a seminar exercise. But I want a job. I want job security. I've got a wife and two kids and another one on the way. I need my geography friends. I want one of those American Geographical Society meritorious awards. I want to be president of the AAG! I don't want any of that crap he's had to put up with.

Cynic.

Realist, you mean. Hey, let's get out of this two-bit whorehouse. I've had enough.

Appendix
Notes
Index

Appendix

Geographers and Others Mentioned in the Text

TO give some "academic" sense of who the people mentioned in the text in one or more of the essays are, I have selected (if available in a search) two published books for each individual, avoiding, as much as possible, joint efforts or ones that are edited. In most but not all cases, I have selected one book written early in the career and another that is more recent, tempered by a sense of "importance" (rather arbitrary) or whether already mentioned in the text. I have tried not to prejudice my selections for those individuals I obviously dislike.

John Adams. Past President of the Association of American Geographers (1982). Professor of Geography, University of Minnesota. Author of *Transport Planning, Vision, and Practice* (1981) and *Housing America in the 1980s* (1987).

John Agnew. Professor and Chair, Department of Geography, University of California at Los Angeles. Author of *Place and Politics: The Geographical Mediation of State and Society* (1987) and *Geopolitics: Re-visioning World Politics* (1998).

James Allen. Professor of Geography, California State University at Northridge. Author of *We the People: An Atlas of America's Ethnic Diversity* (1988) and *The Ethnic Quilt: Population Diversity in Southern California* (with Eugene Turner) (1997).

James Blaut. Deceased. Professor of Geography, University of Illinois at Chicago. Author of *The National Question* (1987) and *The Colonizer's Model of the World: Geographical Diffusionism and Eurocentric History* (1993).

John Borchert. President of the Association of American Geographers (1968). Regents Professor Emeritus, University of Minnesota. Author of *Minnesota's Changing Geography* (1959) and *America's Northern Heartland* (1987).

R. J. Bromley. Professor of Geography, State University of New York at Albany. Author of *Development and Planning in Ecuador* (1977) and *South American Development: A Geographical Introduction* (with Rosemary Bromley) (1988).

249

Stanley Brunn. Editor of the *Professional Geographer* (1982–1987) and of the *Annals of the Association of American Geographers* (1988–1993). Professor of Geography, University of Kentucky. Author of *Geography and Politics in America* (1974) and *The Geography of Laws and Justice: Spatial Perspectives on the Criminal Justice System* (with Keith Harries) (1978).

William Bunge. Author of *Theoretical Geography* (1962) and *Nuclear War Atlas* (1988).

Anne Buttimer. Professor of Geography, University College at Dublin. Author of *Society and Milieu in the French Geographic Tradition* (1971) and *Geography and the Human Spirit* (1993).

Cutler Cleveland. Associate Professor of Geography, Boston University. Editor of *Energy and Resource Quality: The Ecology of the Economic Process* (with Charles A. S. Hall and Robert Kaufmann) (1986) and *The Development of Ecological Economics* (with Robert Costanza and Charles Perrings) (1997).

Kevin Cox. Professor of Geography, Ohio State University. Author of *Conflict, Power, and Politics in the City: A Geographic View* (1973) and *Location and Public Problems: A Political Geography of the Contemporary World* (1979).

Michael Curry. Professor of Geography, University of California at Los Angeles. Author of *The Work in the World: Geographical Practice and the Written Word* (1996) and *Digital Places: Living with Geographic Information Technologies* (1998).

Michael Dear. Professor of Geography, University of Southern California. Author of *State Apparatus: Structures and Language of Legitimacy* (with Gordon Clark) (1984) and *The Postmodern Urban Condition* (2000).

William Denevan. Carl O. Sauer Professor Emeritus of Geography, University of Wisconsin at Madison. Author of *The Aboriginal Cultural Geography of the Llanos de Mojos of Bolivia* (1966) and editor of *The Native Population of the Americas in 1492* (1976).

Debra Dixon. Associate Professor of Geography, East Carolina University.

Daniel Doeppers. Professor of Geography, University of Wisconsin at Madison. Author of *Ethnicity and Class in the Structure of Philippine Cities* (1971) and *Manila, 1900–1941: Social Change in a Late Colonial Metropolis* (1984).

Robin Doughty. Professor of Geography, University of Texas at Austin. Author of *Feather Fashions and Bird Preservation: A Study in Nature Protection* (1975) and *At Home in Texas: Early Views of the Land* (1987).

James Duncan. Fellow, Emmanuel College, Cambridge University. Author of *The City as Text: The Politics of Landscape Interpretation in the Kandyan Kingdom* (1990) and editor of *Place/Culture/Representation* (with David Ley) (1993).

Carville Earle. Editor of the *Annals of the Association of American Geographers* (1994–1996). Carl O. Sauer Professor of Geography, Louisiana State University. Author of *The Evolution of a Tidewater Settlement System: All Hallow's*

Parish, Maryland, 1650–1783 (1975) and *Geographical Inquiry and American Historical Problems* (1992).

Kim England. Associate Professor of Geography, University of Washington. Editor of *Who Will Mind the Baby? Geographies of Child Care and Working Mothers* (1996).

Paul English. Deceased. Former Chair and Professor of Geography, University of Texas at Austin. Author of *City and Village in Iran: Settlement and Economy in the Kirman Basin* (1966) and *World Regional Geography: A Question of Place* (with James Miller) (1977).

Nicholas Entrikin. Former Chair and Professor of Geography, University of California at Los Angeles. Author of *Reflections on Richard Hartshorne's "The Nature of Geography"* (with Stanley Brunn) (1989) and *The Betweenness of Place: Towards a Geography of Modernity* (1991).

John B. Garver, Jr. Ph.D. student of D. W. Meinig. Former Assistant Editor and Chief Cartographer, National Geographic Society. Author of *America's Federal Lands* (map) (with Richard Darley and John Shupe) (1982) and *Deep South* (map) (with Richard Darley and John Shupe) (1983).

Art Getis. Former Head, Department of Geography, University of Illinois at Champaign-Urbana. Professor and Birch Foundation Endowed Chair in Geographic Studies, San Diego State University. Author of *Point Pattern Analysis* (with Barry Boots) (1988) and *Human Geography: Landscapes of Human Activities* (with Jerome Fellmann and Judith Getis) (1990).

Peter Gould. Deceased. Evan Pugh Professor Emeritus of Geography, Pennsylvania State University. Author of *Fire in the Rain* (1990) and *Becoming a Geographer* (1999).

Derek Gregory. Professor of Geography, University of British Columbia. Author of *Ideology, Science, and Human Geography* (1978) and *Geographical Imaginations* (1994).

Susan Hanson. President of the Association of American Geographers (1990). Coeditor of the *Annals of the Association of American Geographers* (1985–1987). Coeditor of *Economic Geography*. Professor of Geography, Clark University. Editor of *The Geography of Urban Transportation* (1986) and author of *Gender, Work, and Space* (with Geraldine Pratt) (1995).

John Fraser Hart. Executive Director of the Association of American Geographers (1965). Editor of the *Annals of the Association of American Geographers* (1970–1975). President of the Association of American Geographers (1979). Professor of Geography, University of Minnesota. Author of *The British Moorlands: A Problem in Land Utilization* (1955) and *The Rural Landscape* (1998).

David Harvey. Distinguished Professor of Anthropology, City University of New York Graduate School. Author of *Explanation in Geography* (1970) and *The Condition of Postmodernity: An Enquiry into the Origins of Cultural Change* (1990).

Kingsley Haynes. Professor of Public Policy, George Mason University. Author of *Gravity and Spatial Interaction Models* (with A. Stewart Fotheringham) (1984) and *Flying into the Future: Air Transport Policy in the European Union* (with Kenneth Button and Roger Stough) (1998).

George Hoffman. Deceased. Professor Emeritus of Geography, University of Texas at Austin. Author of *The Balkans in Transition* (1963) and editor of *Federalism and Regional Development: Case Studies on the Experience in the United States and the Federal Republic of Germany* (1981).

Robert Holz. Former Chair and Erich W. Zimmerman Regents Professor of Geography, University of Texas at Austin. Author of *Economic and Population Growth in the Guadalupe-Blanco River Area* (with Charles Clark) (1971) and editor of *The Surveillant Science: Remote Sensing of the Environment* (1985).

John Hudson. Editor of the *Annals of the Association of American Geographers* (1976–1981). Professor of Geography, Northwestern University. Author of *Geographical Diffusion Theory* (1972) and *Plains Country Towns* (1985).

J. B. Jackson. Deceased. Editor of *Landscape* for seventeen years. Author of *The Necessity for Ruins, and Other Topics* (1980) and *Discovering the Vernacular Landscape* (1984).

John Jakle. Professor of Geography, University of Illinois at Champaign-Urbana. Author of *Images of the Ohio Valley: A Historical Geography of Travel, 1740 to 1860* (1977) and *Fast Food: Roadside Restaurants in the Automobile Age* (with Keith Sculle) (1999).

Preston E. James. Deceased. President and Honorary President of the Association of American Geographers (1951 and 1966). Maxwell Professor of Geography, Syracuse University. Editor of *American Geography: Inventory and Prospect* (with Clarence F. Jones, 1954) and *All Possible Worlds: A History of Geographical Ideas* (with Geoffrey J. Martin, 1981).

Carl Johannessen. Professor Emeritus of Geography, University of Oregon. Author of *Savannas of Interior Honduras* (1963).

Cindi Katz. Professor of Psychology, City University of New York Graduate School. Editor of *Full Circles: Geographies of Women over the Life Course* (with Janice Monk) (1993).

Miriam Kerndt. Geography Librarian, University of Wisconsin at Madison.

C. Gregory Knight. Professor of Geography, Pennsylvania State University. Author of *Ecology and Change: Rural Modernization in an African Community* (1974) and editor of *Contemporary Africa: Geography and Change* (with James Newman) (1976).

Audrey Kobayashi. Professor of Geography, Queen's University. Editor of *Remaking Human Geography* (with Suzanne Mackenzie) (1989) and *Women, Work, and Place* (1994).

Henri Lefebvre. Author of *La Révolution urbaine* (1970) and *The Production of Space* (translated by Donald Nicholson-Smith) (1991).

Peirce Lewis. President of the Association of American Geographers (1983). Professor Emeritus of Geography, Pennsylvania State University. Author of *New Orleans: The Making of an Urban Landscape* (1976).

Ross MacKinnon. Dean, College of Liberal Arts and Sciences, University of Connecticut. Author of *The Concept of Flexibility and Urban Systems* (1968) and *A New Approach to Network Generation and Map Representation: The Linear Case of the Location Allocation Problem* (with Gerry Barber) (1971).

Ian Manners. Professor of Geography, University of Texas at Austin. Author of *The Coastal Energy Impact Program in Texas* (with Wyatt Dietrich and Teri Keen) (1980) and *North Sea Oil and Environmental Planning: The United Kingdom Experience* (1982).

Robert McNee. Deceased. Former Executive Director, American Geographical Society. Professor Emeritus of Geography, University of Cincinnati. Author of *A Primer on Economic Geography* (1971).

Patricia McWethy. Executive Director of the Association of American Geographers (1979–1984).

D. W. Meinig. Maxwell Research Professor of Geography, Syracuse University. Author of *The Great Columbia Plain: A Historical Geography, 1805–1910* (1968) and *Imperial Texas: An Interpretive Essay in Cultural Geography* (1969).

Andy Merrifield. Professor of Geography, Clark University. Editor of *The Urbanization of Injustice* (with Erik Swyngedouw) (1996).

Don Mitchell. Professor of Geography, Syracuse University. Author of *The Lie of the Land: Migrant Workers and the California Landscape* (1996) and *Cultural Geography: A Critical Introduction* (2000).

Richard Morrill. President of the Association of American Geographers (1981). Professor Emeritus of Geography, University of Washington. Author of *The Geography of Poverty in the United States* (1971) and *Teaching Values in College* (1980).

Bernard Nietschmann. Deceased. Professor of Geography, University of California at Berkeley. Author of *Between Land and Water: The Subsistence Ecology of the Miskito Indians, Eastern Nicaragua* (1973) and *Caribbean Edge: The Coming of Modern Times to Isolated People and Wildlife* (1979).

John O'Laughlin. Professor of Geography, University of Colorado. Editor of *Foreign Minorities in Continental European Cities* (with Gunther Glebe) (1987).

Risa Palm. President of the Association of American Geographers (1984). Professor of Geography and Dean, University of Oregon. Editor of *An Invitation to Geography* (with David Lanegran) (1973) and author of *After a California Earthquake: Attitude and Behavior Change* (with Michael Hodgson) (1992).

James Parsons. Deceased. President of the Association of American Geographers (1974). Professor of Geography, University of California at Berkeley. Author of *Antioqueno Colonization in Western Colombia* (1949) and *The Green Turtle and Man* (1962).

Richard Peet. Former Editor of *Antipode: A Radical Journal of Geography.* Coeditor of *Economic Geography.* Professor of Geography, Clark University. Author of *Global Capitalism: Theories of Societal Development* (1991) and *Theories of Development* (with Elaine Hartwick) (1999).

John Pickard. Former Senior Lecturer, Graduate School of the Environment, Macquarie University (Sydney, Australia). Program Manager, New South Wales Parks and Wildlife. Editor of *Antarctic Oasis: Terrestrial Environments and History of the Vestfold Hills* (1986).

Allan Pred. Professor of Geography, University of California at Berkeley. Author of *City Systems in Advanced Economies: Past Growth, Present Processes, and Future Developments* (1977) and *Making Histories and Constructing Human Geographies: The Local Transformation of Practice, Power Relations, and Consciousness* (1990).

Patricia Price. Assistant Professor, Florida International University.

Edward Relph. Professor of Geography, University of Toronto. Author of *Place and Placelessness* (1976) and *The Modern Urban Landscape* (1987).

Bruce Ryan. Former Chair and Professor of Geography, University of Cincinnati.

Robert Sack. Clarence J. Glacken and Bascom Professor of Geography, University of Wisconsin at Madison. Author of *Conceptions of Space in Social Thought: A Geographic Perspective* (1980) and *Homo Geographicus: A Framework for Action, Awareness, and Moral Concern* (1997).

Carl Sauer. Deceased. President and Honorary President of the Association of American Geographers (1940 and 1956). Professor of Geography, University of California at Berkeley. Author of *Agricultural Origins and Dispersals* (1952) and *The Early Spanish Main* (1966).

Rob Shields. Professor of Geography, Carleton University. Author of *Places on the Margin: Alternative Geographies of Modernity* (1992) and *Lefebvre, Love, and Struggle: Spatial Dialectics* (1999).

Neil Smith. Distinguished Professor of Anthropology, City University of New York Graduate School. Author of *Uneven Development: Nature, Capital, and the Production of Space* (1984) and *The New Urban Frontier: Gentrification and the Revanchist City* (1996).

Edward J. Soja. Professor of Geography and Urban Planning, University of California at Los Angeles. Author of *Postmodern Geographies: The Reassertion of Space in Critical Social Theory* (1989) and *Postmetropolis: Critical Studies of Cities and Regions* (2000).

David Sopher. Deceased. Professor of Geography, Syracuse University. Author of *The Sea Nomads: A Study Based on the Literature of the Maritime Boat People of Southeast Asia* (1965) and *Geography of Religions* (1967).

Joseph Spencer. Deceased. Editor of the *Annals of the Association of American Geographers* (1964–1969). Professor of Geography, University of California at Los

Angeles. Author of *Asia, East by South: A Cultural Geography* (1954) and *Oriental Asia: Themes Toward a Geography* (1973).

Howard Stafford. Chair and Professor of Geography, University of Cincinnati. Author of *Planning for Locational Change in the Delivery of Medical Care: A Selected Bibliography* (with Kenneth Corey) (1969) and editor of *Industrial Location and Regional Systems: Spatial Organization in the Economic Sector* (with John Rees and Geoffrey Hewings) (1981).

Paul Starrs. Editor of the *Geographical Review.* Associate Professor of Geography, University of Nevada at Reno. Author of *San Francisco Bay: Its History and Development in Maps* (1993) and *Let the Cowboy Ride: Cattle Ranching in the American West* (1998).

Yi-Fu Tuan. J. K. Wright and Vilas Professor Emeritus of Geography, University of Wisconsin at Madison. Author of *Dominance and Affection: The Making of Pets* (1984) and *Cosmos and Hearth: A Cosmopolite's Viewpoint* (1996).

Richard Walker. Former Chair and Professor of Geography, University of California at Berkeley. Author of *The Capitalist Imperative: Territory, Technology, and Industrial Growth* (with Michael Storper) (1989) and *The New Social Economy: Reworking the Division of Labor* (with Andrew Sayer) (1992).

Michael Webber. Professor, Melbourne University. Author of *Explanation, Prediction, and Planning: The Lowry Model* (1984) and *The Golden Age Illusion: Rethinking Postwar Capitalism* (with David Rigby) (1996).

Wilbur Zelinsky. President of the Association of American Geographers (1972). Professor Emeritus of Geography, Pennsylvania State University. Author of *The Cultural Geography of the United States* (1973) and *Exploring the Beloved Country: Geographic Forays into American Society and Culture* (1994).

Notes

Introduction: The Critic and Criticism

1. Richard Symanski and John A. Agnew, *Order and Skepticism: Human Geography and the Dialectic of Science* (Washington, D.C.: Association of American Geographers, 1981).

2. Richard Symanski, *The Immoral Landscape: Female Prostitution in Western Societies* (Toronto: Butterworths, 1981); *Wild Horses and Sacred Cows* (Flagstaff, Ariz.: Northland Press, 1985); *Outback Rambling* (Tucson: Univ. of Arizona Press, 1990); *Blackhearts: Ecology in Outback Australia* (New Haven: Yale Univ. Press, 2000). The half-page run of dialogue, faithfully recorded, contained several uses of the word *cunt* by an outback Australian. *Cunt* is used as often, if not more so, in the Australian outback by males than is the word *fuck* by certain blue-collar males in the United States. The fifteen hundred-word section was a heavily documented essay dealing with a proposed Jewish settlement in 1939 on the cattle station where the research of interest took place. Apparently, and in ways never specified for me, it struck someone (whom, I was never told) as "anti-Semitic"; six scholars I had examine the essay could find nothing remotely anti-Semitic in the piece.

3. I am categorically opposed to anonymous reviews. See Richard Symanski and John Pickard, "Rules by which We Judge One Another," *Progress in Human Geography* 20 (1996): 175–82.

Calling a Spade a Spade

1. D. W. Meinig, *Imperial Texas* (Austin: Univ. of Texas Press, 1969); Richard Symanski, "The Manipulation of Ordinary Language," *Annals of the Association of American Geographers* 66 (1976): 605–14.

2. Richard Symanski, "God, Food, and Consumers in Periodic Market Systems," *Proceedings of the Association of American Geographers* 5 (1973): 262–66; Michael J. Webber and Richard Symanski, "Periodic Markets: An Economic Location Analysis," *Economic Geography* 49 (July 1973): 213–27; Richard Symanski and Michael J. Webber, "Complex Periodic Market Cycles," *Annals of the Association of American Geographers* 64 (June 1974): 203–13; R. J. Bromley and Richard Symanski, "Marketplace Trade in Latin

257

America," *Latin American Research Review* 9 (fall 1974): 3–38; Richard Symanski and R. J. Bromley, "Market Development and the Ecological Complex," *Professional Geographer* 26 (Nov. 1974): 382–88; Michael J. Webber, Richard Symanski, and James Root, "Toward a Cognitive Spatial Theory," *Economic Geography* 51 (Apr. 1975): 100–116; Richard Symanski, Ian Manners, and R. J. Bromley, "The Mobile-Sedentary Continuum," *Annals of the Association of American Geographer* 65 (Sept. 1975): 461–71; A. Norton and Richard Symanski, "The Internal Marketing Systems of Jamaica," *Geographical Review* 65 (Oct. 1975): 461–75; R. J. Bromley, Richard Symanski, and C. M. Good, "The Rationale of Periodic Markets," *Annals of the Association of American Geographers* 65 (Dec. 1975): 530–37; Richard Symanski, "How Periodic Markets Determine Their Major Market Day," in *American Philosophical Society Yearbook, 1975* (Philadelphia: American Philosophical Society, 1976), 478–79; Richard Symanski and Michael J. Webber, "Large Cities and Market Day Patterns," in *Market Distribution Systems*, edited by Erdmann Gormsen (Mainz, Germany: Geographisches Institut der Johannes-Guttenberg-Universitat, 1976), 17–25; Richard Symanski, "Periodic Markets in Southern Colombia," in *MarketPlace Trade: Periodic Markets, Hawkers, and Traders in Africa, Asia, and Latin America*, edited by Robert H. T. Smith (Vancouver, B.C.: Centre for Transportation Studies, Univ. of British Columbia, 1978), 171–85.

3. Richard Symanski, "Prostitution in Nevada," *Annals of the Association of American Geographers* 64 (1974): 357–77.

4. Symanski, "The Manipulation of Ordinary Language," 609.

5. David Sopher, "Commentary: Ordinary Language," *Annals of the Association of American Geographers* 67 (1977): 625–26; G. R. Vale, "Ordinary Language," *Annals of the Association of American Geographers* 67 (1977): 626–29; O. F. G. Sitwell, "Did Meinig Tell the Truth?" *Annals of the Association of American Geographers* 67 (1977): 629–32; Richard Symanski, "Comment in Reply," *Annals of the Association of American Geographers* 67 (1977): 633–39.

Deconstructing a Drugstore Cowboy

1. William M. Denevan, ed., *Hispanic Lands and Peoples: Selected Writings of James J. Parsons*, Dellplain Latin American Studies, no. 23 (Boulder: Westview Press, 1989).

2. Paul F. Starrs, *Let the Cowboy Ride: Cattle Ranching in the American West* (Baltimore: Johns Hopkins Univ. Press, 1998).

3. Ibid., 343. Kendall L. Johnson, *Rangeland Through Time: A Photographic Study of Vegetation Change in Wyoming, 1870–1896*, Agricultural Experiment misc. pub. 50 (Laramie: University of Wyoming, 1987); Gary F. Rogers, *Then and Now: A Photographic History of Vegetation Change in the Central Great Basin Desert* (Salt Lake City: University of Utah Press, 1982).

4. Starrs, *Let the Cowboy Ride*, 344. Denzel Ferguson and Nancy Ferguson, *Sacred Cows at the Public Trough* (Bend, Oreg.: Maverick Publications, 1983).

5. Starrs, *Let the Cowboy Ride*, 344. Lynn R. Jacobs, *Waste of the West: Public Lands Ranching* (Tucson: by the author, 1991).

6. Starrs, *Let the Cowboy Ride*, 344. Jeremy Rifkin, *Beyond Beef: The Rise and Fall of the Cattle Culture* (New York: Dutton, 1992).

7. Debra L. Donahue, *The Western Range Revisited: Removing Livestock from Public Lands to Conserve Native Biodiversity* (Norman: Univ. of Oklahoma Press, 1999).

8. Starrs, *Let the Cowboy Ride*, 317.

9. Ibid., 21.

10. Ibid., 67.

11. Ibid., 200; italics added.

12. Ibid., 255.

13. Http://www.geography.unr.edu/geohome/faculty/. For a more detailed discussion of Starrs's storytelling about himself and his meaningless prose, see Korski, "Hugs and Kisses All Around: The Berkeley Boys and Father Tuan" (available from author upon request).

Academic Brahmins

1. Edward W. Soja, *Thirdspace: Journeys to Los Angeles and Other Real-and-Imagined Places* (Cambridge, Mass.: Blackwell, 1996); Debra Dixon, "Between Difference and Alterity: Engagements with Edward Soja's *Thirdspace*," *Annals of the Association of American Geographers* 89 (1999): 338–39; Andy Merrifield, "The Extraordinary Voyages of Ed Soja: Inside the 'Trialectics of Spatiality,' " *Annals of the Association of American Geographers* 89 (1999): 345–48; Patricia Price, "Longing for Less of the Same," *Annals of the Association of American Geographers* 89 (1999): 342–44; Rob Shields, "Harmony in Thirds: Chora for Lefebvre," *Annals of the Association of American Geographers* 89 (1999): 340–42; Edward Soja, "Keeping Space Open," *Annals of the Association of American Geographers* 89 (1999): 348–52.

2. Edward Soja, *The Geography of Modernization in Kenya: A Spatial Analysis of Social, Economic, and Political Change* (Syracuse: Syracuse Univ. Press, 1968).

3. Edward Soja, *Postmodern Geographies: The Reassertion of Space in Critical Social Theory* (London: Verso, 1989).

4. Dixon, "Between Difference and Alterity," 338; italics added.

5. Price, "Longing for Less," 342.

6. Merrifield, "Extraordinary Voyages," 345.

7. Ibid.

8. Soja, *Thirdspace*, 349.

9. Shields, "Harmony in Thirds," 341.

10. Soja, *Thirdspace*, 349.

11. Price, "Longing for Less," 343.

12. Soja, *Thirdspace*, 349, 350.

13. Ibid., 351.

14. Ibid., 353.

By My Sins Shall I Be Redeemed

1. Jim Duncan, "Complicity and Resistance in the Colonial Archive: Some Issues of Method and Theory in Historical Geography," *Historical Geography* 27 (1999): 119–28.

2. Ibid., 127.

3. Ibid., 125.

4. Ibid., 124.

5. See G. Obeyesekere, *The Apotheosis of Captain Cook: European Mythmaking in the Pacific* (Princeton: Princeton Univ. Press, 1992). The lousy scholarship and misplaced charges by Obeyesekere are nicely countered in M. Sahlins, *How "Natives" Think about Captain Cook, for Example* (Chicago: Univ. of Chicago Press, 1995).

6. Duncan, "Complicity and Resistance," 122.

7. Ibid., 123.

When Logic and Language Go on a Walkabout

1. Robert David Sack, *Homo Geographicus: A Framework for Action, Awareness, and Moral Concern* (Baltimore: Johns Hopkins Univ. Press, 1997).

2. For more on Yi-Fu Tuan and why he is everything to which geographers should not want to aspire, see Korski, "Hugs and Kisses All Around."

3. For a critique of Sack's prize Ph.D. student, Nick Entrikin, another example of vapid geography, see Korski, "The Landscape of Narrative-Like Syntheses" (available from author upon request).

4. Sack, *Homo Geographicus*, 132. Unless otherwise indicated, other quotes in this section are from the same page.

5. Ibid., 132–33.

6. Karl S. Zimmerer, *Changing Fortunes: Biodiversity and Peasant Livelihood in the Peruvian Andes* (Berkeley and Los Angeles: Univ. of California Press, 1996).

7. Sack, *Homo Geographicus*, 85.

8. Ibid., 95.

9. Ibid., 60.

10. Ibid., 31.

11. Robert David Sack, *Place, Modernity, and the Consumer's World: A Relational Framework for Geographical Analysis* (Baltimore: Johns Hopkins Univ. Press, 1992).

Words in the World, Worthless Words

1. Michael R. Curry, *The Work in the World: Geographical Practice and the Written Word* (Minneapolis: Univ. of Minnesota Press, 1996), ix.

2. Symanski, "Manipulation of Ordinary Language"; Richard Symanski, "Why We Should Fear Postmodernists," *Annals of the Association of American Geographers* 85 (1994): 301–4.

3. Notable in this regard is the work of Bret Wallach, who can consistently turn a beautiful phrase or sentence, yet often seems to be utterly bereft of analytic ability—and

often just plain lost. See *At Odds with Progress: Americans and Conservation* (Tucson: Univ. of Arizona Press, 1991) and *Losing Asia: Modernization and the Culture of Development* (Baltimore: Johns Hopkins Univ. Press, 1996).

4. Symanski, "Manipulation of Ordinary Language"; Symanski, "Why We Should Fear Postmodernists."

5. Richard Rorty, *Objectivity, Relativism, and Truth* (Cambridge: Cambridge Univ. Press, 1991).

6. For example, and for openers, there's a good deal that he could have done with the tough, analytic work of Stanley Fish in this regard. See *Is There a Text in This Class? The Authority of Interpretative Communities* (Cambridge: Harvard Univ. Press, 1980) and *Doing What Comes Naturally: Change, Rhetoric, and the Practice of Theory in Literary and Legal Studies* (Durham: Duke Univ. Press, 1989). Also see R. Kimball, *Tenured Radicals: How Politics Has Corrupted Our Higher Education* (New York: Harper and Row, 1990), 166–89.

7. Henri Lefebvre, *The Production of Space,* translated by Donald Nicholson-Smith (Oxford, England, and Cambridge, Mass.: Blackwell, 1991).

8. Curry, *Work in the World,* 194–95.

Ruminations on a Misshapen America

1. D. W. Meinig, *The Shaping of America: A Geographical Perspective on 500 Years of History, Volume I, Atlantic America, 1492–1800* (New Haven: Yale Univ. Press, 1986); *Volume II, Continental America, 1800–1867* (New Haven: Yale Univ. Press, 1993); *Volume III, Transcontinental America, 1850–1915* (New Haven: Yale Univ. Press, 1998).

2. Symanski, "Manipulation of Ordinary Language."

3. David S. Landes, *The Wealth and Poverty of Nations: Why Some Are So Rich and Some So Poor* (New York: W. W. Norton, 1999).

4. Meinig, *Shaping of America,* 1:7.

5. Ibid., 65.

6. Ibid., 258, 261–62.

7. Ibid., 2:176.

8. Ibid., 45–46.

9. Ibid., xviii.

10. Ibid., 130.

11. Ibid., 231.

12. William Bunge, *Theoretical Geography* (Lund, Sweden: C. W. K. Gleerup, 1962).

13. Meinig, *Shaping of America,* 1:xvi.

14. Ibid., 45, 126, 212, 30, 178 (last two quotes).

15. Ibid., 8, 107, 145, 101, 206–7; italics added.

16. Ibid., 9, 51, 213; italics added.

17. I had an excellent editor for my most recent book, *Blackhearts.* Although I by no means took all of her suggestions, the many I did undoubtedly improved tone and readability.

18. Even if Meinig was exaggerating, I'm prepared to believe that he got little help,

and that it was due in no small way to his reputation and sense of self-importance. The source of this information is reliable, but the informant, as with others who gave me advice on some of the essays in this book, is not eager to be identified. There is never a shortage of "Deep Throats."

Coconuts on a Lava Flow in the Chiricahua Mountains

1. For a debunking of Carl Sauer, the man, the scholar, and the legend, see Korski, "Looking for the Real Carl Sauer" (available from author upon request).

2. Bret Wallach, "Will Carl O. Make It Across That Big Ol' Bridge That Bill and Al Keep Talking About?" a version of a paper published in *Yearbook of the Association of Pacific Coast Geographers* 61 (1999): 1; see http://geography.ou.edu/research/paper.html.

3. Bob Callahan, introduction to *Carl O. Sauer: Selected Essays, 1963–1975* (Berkeley: Turtle Island Foundation, 1981), xv. Daniel W. Gade, a cultural geographer long interested in crops, claimed that "Sauer's influence and that of his own students account for the content of cultural geography over the last half century" *(Nature and Culture in the Andes* [Madison: Univ. of Wisconsin Press, 1999], 21). Glacken's book is *Traces on the Rhodian Shore: Nature and Culture in Western Thought from Ancient Times to the End of the Eighteenth Century* (Berkeley and Los Angeles: Univ. of California Press, 1967).

4. Joe Spencer to Carl Sauer, Sept. 20, 1943, Sauer Papers, Bancroft Library, University of California at Berkeley.

5. David Stoddart, "Carl Sauer: Geomorphologist," in *Process and Form in Geomorphology* (London: Routledge, 1997), 340–79. Sauer was still very much interested in doing geomorphology as late as 1948. In that year, he was eager to join a team that was going to Mexico, saying that he wanted to record the presence or absence of marine terraces, evidence of vertebrate and invertebrate remains, and evidence of human occupance on the terraces (Carl Sauer to J. W. Sefron Jr., Dec. 18, 1948, Sauer Papers).

6. Stoddart, "Carl Sauer: Geomorphologist," 357. The study in question is Carl Sauer, *Land Forms in the Peninsular Range of California As Developed about Warner's Hot Springs and Mesa Grande,* University of California Publications in Geography, vol. 3, no. 4 (Berkeley and Los Angeles: Univ. of California Press, 1929), 199–290.

7. Stoddart, "Carl Sauer: Geomorphologist," 357. There were other instances where Sauer simply refused to answer his critics, as when E. D. Merrill attacked him for his competence as a biogeographer and what he had to say in *Agricultural Origins and Dispersals* (Henry Bruman, "Carl Sauer in Mid-Career: A Personal View by One of His Students," in *Carl O. Sauer: A Tribute,* edited by Martin Kenzer [Corvallis: Univ. of Oregon Press, for the Association of Pacific Coast Geographers, 1987], 132. That the remarks by Merrill may have been intemperate, as Bruman claims, is not reason for not engaging a critic.

8. Stoddart, "Carl Sauer: Geomorphologist," 352. Several decades later, Sauer would write up his ideas about the importance of the seashore to Early Man ("Seashore: Primitive Home of Man?" in *Land and Life: A Selection from the Writings of Carl Ortwin Sauer,* edited by John Leighly [Berkeley and Los Angeles: Univ. of California, 1967], 300–312.

9. Stoddart, "Carl Sauer: Geomorphologist," 359; italics added. The study in question is Carl Sauer, *Basin and Range Forms in the Chiricahua Area*, University of California Publications in Geography, vol. 3, no. 6 (Berkeley and Los Angeles: Univ. of California Press, 1930), 339–414.

10. Stoddart, "Carl Sauer: Geomorphologist," 360.

11. Robert C. West, ed., *Andean Reflections: Letters from Carl O. Sauer While on a South American Trip Under a Grant from the Rockefeller Foundation, 1942* (Boulder: Westview Press, in cooperation with the Department of Geography, Syracuse Univ., Dellplain Latin American Studies, no. 11, 1982), 25, 27–28. All this railway-car speculation, alas, reminds me of the wrongheaded and pernicious generalizations that Paul Theroux has so often been quick to make about whole ethnic groups, while looking through the window from his train seat, making up the world as it flashes by. Nothing better illustrates this kind of writing than all the unnerving silliness we get in his *Old Patagonian Express: By Train Through the Americas* (Boston: Houghton Mifflin, 1979).

12. Stoddart, "Carl Sauer: Geomorphologist," 364, 394.

13. Peter Gould, "Geography, 1957–1977: The Augean Period," in *Becoming a Geographer* (Syracuse: Syracuse Univ. Press, 1999), 86. The full quotation reads as follows: "The Polacca Wash is a place of evil omen. Perhaps you can turn it to a more fortunate connotation than it has had thus far. One of the difficulties with it has been that personal jealousies have flared very sharply in the work that has been done on it" (Carl Sauer to Earl F. Dosch, Nov. 4, 1937, Sauer Papers).

14. Robert C. West, *Carl Sauer's Fieldwork in Latin America* (Syracuse: Syracuse Univ., Department of Geography, 1979), 35–36. Sauer could get quite testy about criticism and the manner in which it was carried out if his findings or ideas were at stake. When he heard about an article that had repudiated by some archaeological work in Tepexpan, Mexico, with which he had been involved, he complained that it was the duty of the people repudiating what had been found to communicate directly with the person with whom they were in disagreement. "If the visitors then found that the grateful owners of the problem were obdurate and insisted upon putting a wrong face on the matter, they should later express and document dissent. I am not concerned simply with professional ethics. I am interested in seeing the evidence fully worked up and fairly presented for all its worth. The main thing I fear is premature throwing of things out of court" (Carl Sauer to Paul Sears, Oct. 22, 1948; and Paul Sears to Carl Sauer, Oct. 6, 1948, Sauer Papers). Sauer's position is honorable indeed, but whether he behaved similarly is another matter.

15. West, *Sauer's Fieldwork in Latin America*, 12–13, 34.

16. Henry J. Bruman, "Some Observations on the Early History of the Coconut in the New World," *Acta Americana* 2 (1944): 220–43; Henry J. Bruman, "Early Coconut Culture in Western Mexico," *Hispanic American Historical Review* 25 (1945): 212–23; Henry J. Bruman, "A Further Note on Coconuts in Colima," *Hispanic American Historical Review* 27 (1947): 572–73; Carl O. Sauer, "Cultivated Plants of South and Central America," in vol. 6 of *Handbook of South American Indians*, edited by Julian H. Steward, Smithsonian Institution, Bureau of American Ethnology Bulletin 143 (Washington, D.C.: U.S. Government Printing Office, 1950), 487–543; Bruman, "Carl Sauer in Mid-Career," 132.

17. Henry Bruman to Carl Sauer, May 6, 1944, Sauer Papers.

18. Carl Sauer to Henry Bruman, May 1944, Sauer Papers. There are cases where Sauer seemed to think that being dogmatic was quite okay. In an exchange with Glenn Trewartha, in which Sauer was trying hard to get a German academic by the name of Credner hired at the University of Wisconsin, he wrote the following: "Is his dogmatic quality basic, or is it a device by which you stir up discussion, dissent, and prepare the way for agreement? Credner lived a good deal with Waibel and Waibel used that device very effectively. I have had some terrific scraps with Waibel but found an aftereffect of strong stimulation!" (Nov. 14, 1938, Sauer Papers). Is this how Sauer felt about himself, or was he in fact just plain dogmatic about most of his ideas?

19. Sauer to Bruman, May 1944, Sauer Papers. Carl Sauer, *Agricultural Origins and Dispersals* (New York: American Geographical Society, 1952).

20. Henry Bruman to Carl Sauer, June 7, 1944, Sauer Papers. There were apparently a few exceptions. For example, in a 1960 letter to Raymond Dart, Sauer wrote, "At one time I discounted the seaside habitat as germinal because of the extreme conservatism of life. I should do so no longer but join with you in attributing to it at one time the leading role as kindergarten of culture. I think however the greatest discovery of Early Man was when he wandered down to some tidal seashore there to fine inexhaustible abundance of food and raw materials" (June 18, 1960, Sauer Papers).

21. Carl Sauer to Henry Bruman, June 13, 1944, Sauer Papers.

22. Marvin Mikesell, on the other hand, is keen to rebut Sauer's "low-tech, humanistic" image and replace it with what he calls his "scientific career" ("Sauer and 'Sauerology': A Student's Perspective," in *Carl O. Sauer: A Tribute*, edited by Kenzer, 146.

23. Gade, in *Nature and Culture* (197), would simply reduce the problem to Sauer putting too much emphasis on the work of O. F. Cook ("History of the Coconut Palm in America," *Contributions of the U.S. National Herbarium* 12 [1910]: 271–342).

24. Bruman, "Carl Sauer in Mid-Career," 132.

25. Quote from Gade, *Nature and Culture*, 197. B. K. Maloney, "Paleoecology and the Origins of the Coconut," *GeoJournal* 31 (1993): 355–62; Jonathan Sauer, *Historical Geography of Crop Plants: A Select Roster* (Boca Raton: CRC Press, 1993), 186–89. On the issue of *Canavalia*, Gade notes that Jonathan Sauer disapproved of his father's idea that humans carried this legume across oceans. To add heft to the argument, he put fifty-one species of *Canavalia* into an evolutionary framework and then reconstructed their histories and distributions (*Nature and Culture*, 201).

26. J. Harlan, "Plant Domestication: Diffuse Origins and Diffusion," in *The Origin and Domestication of Cultivated Plants*, edited by C. Barigozzi (New York: Elsevier, 1986), 21–34. Also see Gade, *Nature and Culture*, 201. Gade claims that Sauer's problem here resulted from an overreaction against ecological theorizing (17). This may be true, but again it may be only a part of the problem, and a more serious one might have been the way his mind closed down and was not open to criticism once he settled on a point of view.

27. Gade, *Nature and Culture*, 195, 202–3. Sauer, for example, misconstrued the range of chayote *(Sechium edule)*, coca was more widely grown than he imagined, capsicum peppers were more varied, and he was wrong about the geographical origins of the

fruit cherimoya. He had intended to make a complete survey of tropical perennials in his 1940s work, but left out more than seventy-five of them. To be sure, it was less important because the number of species known is, in good part, time dependent. Nor were Sauer's taxonomic categories consistent, which he defended by calling what he wrote a "primer for anthropologists" (203).

28. Bruman, "Carl Sauer in Mid-Career," 133, 129. Writing about the experience in 1987, Bruman claimed that Sauer "considered personal vendettas as worse than useless and refused to participate" (132). Whether this claim was entirely true, even in Bruman's case, is open to question. When Bruman's name came up for admission to the Association of American Geographers in 1943, Sauer, who was one of a small group involved with nominating members, said that he could not support Bruman's nomination. To one committee member, he wrote: "I am sorry, therefore, that I cannot at present support a nomination for Henry Bruman. He could have made a scholar of himself of whom we should all have been proud [note the 'could have made']. If he still shows that he has the bent of a scholar, I'll support him. At Henry's present stage of development I think it would be most unfortunate for him to get the idea that scholarship is a matter of indifference and that the ability to promote himself is all that counts" (Carl Sauer to Raymond E. Murphy, Dec. 21, 1943, Sauer Papers). Years later, the "coconut issue" was still a topic of conversation between Sauer and some of his students (see Joe Spencer to Carl Sauer, Nov. 26, 1952; and Carl Sauer to Joe Spencer, Dec. 4, 1952, Sauer Papers). In the latter letter, Sauer seems curiously eager to make little of the coconut. "I doubt that the coconut is very important in the development of cultures. I am rather inclined to look on it as something that was carefully used only in lands unsuited to other starch, sugar, and other fiber plants. The things are picturesque on the atoll, but I think they have been overrated."

29. Gade, *Nature and Culture,* 190.

30. Ibid. See H. Cutler, "Races of Maize in South America," *Botanical Museum Leaflets* (Cambridge, Mass.) 12 (1946): 257–91; J. B. Bird, "Preceramic Cultures in Chicama and Viru," *American Antiquity* 13 (1948): 21–28; W. H. Hodge, "Three Neglected Andean Tubers," *Journal of the New York Botanical Garden* 47 (1946): 214–24; W. La Barre, "Potato Taxonomy among the Aymara Indians of Bolivia," *Acta Americana* 5 (1947): 83–103; R. N. Salaman, *The History and Social Influence of the Potato* (Cambridge: Cambridge Univ. Press, 1949); and M. Cardenas, "Plantas Alimenticias Nativas de los Andes de Bolivia," parts 1–3, *Folia Universitaria* (Cochabamba) 2 (1948): 36–51; 3 (1949): 102–19; 4 (1950): 86–108; Gade, *Nature and Culture,* 190.

31. Late in life, Bruman addressed the coconut issue and how he saw Sauer in light of this decade-long misunderstanding. What is dispiriting about the account is that in spite of Sauer's boorish and unscholarly behavior, Bruman nevertheless described Sauer as a "wonderful man" and "a brilliant scholar" (370, 372). Like so many geographers, and others, he was blinded by Sauer's aura and the questions he addressed. Generously, it can be concluded that Bruman simply did not know what the word scholar means. See Henry J. Bruman, "Recollections of Carl Sauer and Research in Latin America," *Geographical Review* 86 (1996): 370–76.

32. For more on this argument, see Korski, "Looking for the Real Carl Sauer."

33. Ibid.

Lost in Lies about California's Migrant Labor Landscapes

1. Don Mitchell, *The Lie of the Land: Migrant Workers and the California Landscape* (Minneapolis: Univ. of Minnesota Press, 1996), 200.

2. Ibid., 17.

3. Ibid., 18.

4. James J. Parsons, "A Geographer Looks at the San Joaquin Valley," *Geographical Review* 76 (1986): 371–89. The piece was originally presented as the Carl O. Sauer memorial lecture on April 30, 1986, at Berkeley.

5. Mitchell, *Lie of the Land*, 202.

6. James J. Parsons, *Antioqueño Colonization in Western Colombia*, 2d ed., rev. (Berkeley and Los Angeles: Univ. of California Press, 1968). By the mid-1980s, the book had gone through three Spanish editions. Also see James J. Parsons, "Medellin Reconsidered," in *Hispanic Lands and Peoples*, edited by Denevan, 133–40.

7. See Denevan, *Hispanic Lands and Peoples*; and James J. Parsons, *Antioquia's Corridor to the Sea: An Historical Geography of the Settlement of Urabá*, Ibero-Americana, vol. 49 (Berkeley and Los Angeles: Univ. of California Press, 1967).

8. Neil Smith was Don Mitchell's Ph.D. adviser at Rutgers University. See Neil Smith, *Uneven Development: Nature, Capital, and the Production of Space* (New York: Blackwell, 1984). Smith has been one of the great admirers and promoters of Henri Lefebvre, discussed in the essay "Words in the World, Worthless Words."

9. Parsons, "Geographer Looks," 374, 375.

10. Ibid., 375; italics added.

11. Mitchell, *Lie of the Land*, 205.

12. Parsons, "Geographer Looks," 374.

13. Mitchell, *Lie of the Land*, 18 (italics added); Parsons, "Geographer Looks," 378.

14. Parsons, "Geographer Looks," 378.

15. Mitchell, *Lie of the Land*, 28.

Intellectual Turf and Elitist Notions about Invisible Landscapes

1. Richard Symanski, "Good as Gould," *Geographical Review* 90 (2000): 248–55.

2. For a critique of a revengeful book review about Richard Symanski's *Blackhearts*, see Korski, "Robin Lays a Blue Egg: Review of a Black Heart Review of *Blackhearts*" (available upon request from author).

3. Dick Walker, review of *Lie of the Land*, by Mitchell, *Geographical Review* 87 (1997): 408–11. It is worth noting that the same Paul Starrs discussed in the essay "Deconstructing a Drugstore Cowboy," whom I describe as a polemicist and a lousy scholar, described Mitchell's effort in similar terms. He wrote that it's a "fine little book with great pretensions" by someone who bears watching as a "prized polemicist." Further on in the review, Starrs wrote that the book is a kind of "seigneurial socialism that is thrown together more with ideology than with the loving care and cautious weighing and telling that the rigorous scholar musters" (review of *Lie of the Land*, by Mitchell, *Journal of Historical Geography* 25 (1999): 428–30.

4. Walker, review of *Lie of the Land*, 408.

5. Mitchell, *Lie of the Land*, 22.

6. Walker, review of *Lie of the Land*, 408.

7. Ibid., 409; italics added.

8. Ibid., 408. For a very good example of just how people-shy Wilbur Zelinsky was over a period of some forty years, see Wilbur Zelinsky, *Exploring the Beloved Country: Geographic Forays into American Society and Culture* (Iowa City: Univ. of Iowa Press, 1994).

9. Walker, review of *Lie of the Land*, 408.

10. Ibid., 410; italics added.

11. Ibid.

12. Ibid., 411.

Feminist Field-Worker Conceits

1. Heidi J. Nast, "Women in the Field: Critical Feminist Methodologies and Theoretical Perspectives," *Professional Geographer* 46 (1994): 54–66; Cindi Katz, "Playing the Field: Questions of Fieldwork in Geography," 67–72; Audrey Kobayashi, "Coloring the Field: Gender, 'Race,' and the Politics of Fieldwork," 73–80; Kim V. L. England, "Getting Personal: Reflexivity, Positionality, and Feminist Research," 80–89; Melissa R. Gilbert, "The Politics of Location: Doing Feminist Research at 'Home,' " 90–96; and Lynn A. Staeheli and Victoria A. Lawson, "A Discussion of 'Women in the Field': The Politics of Feminist Fieldwork," 96–101.

2. Eric Waddell, *The Mound Builders: Agricultural Practices, Environment, and Society in the Central Highlands of New Guinea* (Seattle: Univ. of Washington Press, 1972); Bernard Nietschmann, *Between Land and Water: The Subsistence Ecology of the Miskito Indians, Eastern Nicaragua* (New York: Seminar Press, 1973); Bernard Nietschmann, *Caribbean Edge: The Coming of Modern Times to Isolated People and Wildlife* (Indianapolis: Bobbs-Merrill, 1979); Bernard Nietschmann, *The Unknown War: The Miskito Nation, Nicaragua, and the United States* (New York: Freedom House; distributed by arrangement with Lanham, Md.: Univ. Press of America, 1989).

3. Katz, "Playing the Field," 72.

4. Ibid., 69.

5. Ibid., 71.

6. England, "Getting Personal," 84.

7. Ibid., 82.

8. Kobayashi, "Coloring the Field," 77.

New Orleans Folks and Fictions

1. Peirce Lewis, *New Orleans: The Making of an Urban Landscape* (Cambridge, Mass.: Ballinger, 1976).

2. See, for example, A. P. Andrus et al., *Seattle* (Cambridge, Mass.: Ballinger, 1976); R. Abler, J. S. Adams, and J. R. Borchert, *The Twin Cities of St. Paul and Minneapolis*

(Cambridge, Mass.: Ballinger, 1976); and M. P. Conzen and G. K. Lewis, *Boston: A Geographical Portrait* (Cambridge, Mass.: Ballinger, 1976).

3. Lewis, *New Orleans*, xiv, xiii (middle three quotes), xiv.

4. Ibid., 115.

5. Ibid., 1.

6. See Jan Morris, *Among the Cities* (New York: Viking, 1985) and *City to City* (Toronto: Macfarlane Walter and Ross, 1990).

7. Lewis, *New Orleans*, 51.

8. In fact, I have written such a book: *Eight-seven Days in the Gentle City: A Personal Journey Through Melbourne*, under review.

9. Lewis, *New Orleans*, 1, xiii.

Arrogant Eyes

1. This essay would have never been written had several eminent geographers and their camp followers been willing to dialogue with me over the years about the issues I raise. But like so many academics—most it seems, in my experience—they find it all too comforting to live within their own little comfy, communal cocoons while at the same time proudly asserting that they're out there on the playing fields where ideas are discussed and debated. Most of the documented analysis in this essay was written in early 1988. Then, no one wanted to publish the substance of this argument because it was too "hot" and would have been an unflattering portrayal of all those cultural geographers in that leaky boat called ordinary landscape analysis (see esp. D. W. Meinig, ed., *The Interpretation of Ordinary Landscapes: Geographical Essays* [New York: Oxford Univ. Press, 1979]). In later years, when I made further attempts to get these ideas published, the claim was that what I was saying was "old hat"—which was false. The initial rejections of this idea by the editor of the *Annals of the Association of American Geographers*, Stanley Brunn, provide disheartening insight into the doublespeak of journal editors I've encountered through the years. In letters exchanged between Brunn and me from Nov. 25 to Dec. 18, 1988, Brunn, in his desire to reject a manuscript favorably reviewed, ran around, under, and over the mulberry bush a dozen different ways: He wanted to reject the manuscript because an abstract was not sent. He wanted numerous points addressed that would have expanded the manuscript to half-again its length when it was already at the limit of acceptable length, and I had to stay within the limit. He refused to send some of the reviews. He refused to consider documented responses showing that key criticisms were just plain wrong. He counted as a review marginal comments on eight pages of a thirty-seven-page manuscript. He took issue with material that was carefully documented in the manuscript, saying that so-and-so just didn't "feel" that way because *he* knew him personally. He suggested that the manuscript be sent to the *Professional Geographer*, which, of necessity, would have cut its length by more than half, or to about a third of what Brunn would have required to meet his objections. And then, when there were issues to be resolved, he set himself up as judge and jury of a topic about which he knew nothing at all. All of this, of course, was nothing more than protection of the power

elite in a major branch of cultural geography at this time and fear of the criticism he, Brunn, would get for publishing a damning critique of an inferior way of knowing ordinary landscapes. Courage has never been a trait to ascribe to geography's journal editors.

2. Gail Sheehy, *Hustling: Prostitution in Our Wide Open Society* (New York: Delacorte Press, 1973).

3. Peirce Lewis, "The Monument and the Bungalow," *Geographical Review* 88 (1998): 507–27.

4. Richard Symanski, "Honduras: When the Saints Arrive," *Geographical Review* 88 (1988): 571–79.

5. Peirce Lewis, "Small Town in Pennsylvania," *Annals of the Association of American Geographers* 62 (1972): 328.

6. J. B. Jackson, *Discovering the Vernacular Landscape* (New Haven: Yale Univ. Press, 1984), 48.

7. Meinig, introduction to *Interpretation of Ordinary Landscapes*, 3.

8. D. W. Meinig, "Environmental Appreciation: Localities as a Humane Art," *Western Humanities Review* 25 (1971): 3.

9. Ibid., 6. Also see D. W. Meinig, "Geography as an Art," *Transactions of the Institute of British Geographers* 8 (1983): 314–28.

10. Meinig, "Environmental Appreciation," 6–7.

11. Peirce F. Lewis, "Axioms for Reading the Landscape," in *Interpretation of Ordinary Landscapes*, edited by Meinig, 11–32.

12. Lewis, "Axioms," 19, 26, 27, 32.

13. D. W. Meinig, "The Beholding Eye: Ten Versions of the Same Scene," in *Interpretation of Ordinary Landscapes*, 33–48; D. W. Meinig, *Southwest: Three Peoples in Geographical Change, 1600–1700* (New York: Oxford Univ. Press, 1971); Meinig, *Imperial Texas*, 91–109, 11.

14. Quoted in Meinig, *Interpretation of Ordinary Landscapes*, 228; "J. B. Jackson's Home Ground," *Landscape Architecture* 78 (1988): 52.

15. J. B. Jackson, "The Accessible Landscape," *Whole Earth Review* 58 (1988): 7; E. H. Zube, ed., *Landscapes: Selected Writings of J. B. Jackson* (Amherst: Univ. of Massachusetts Press, 1970), 88–91.

16. Zube, *Landscapes*, 155; Meinig, "Reading the Landscape," in *Interpretation of Ordinary Landscapes*, 228–29.

17. Zube, *Landscapes*, 86.

18. Nothing illustrates this lack in Zelinsky's case better than his lifelong collection of essays on the United States: dry, trivial, questionable as to conclusions, and a monumental testament to hundreds of hours of driving everywhere and taking reams of notes, but nary a one, it would seem, that came from talking to anyone but himself—unless it was about how logs had been split in the construction of Georgia log cabins (see Zelinsky, *Exploring the Beloved Country*).

19. Edward Abbey, *Desert Solitaire: A Season in the Wilderness* (New York: Simon and Schuster, 1968); William Least Heat Moon, *A Journey into America: Blue Highways* (Boston: Little, Brown, 1982); Barry Lopez, *Arctic Dreams: Imagination and Desire in a*

Northern Landscape (New York: Charles Scribner's Sons, 1986); Barry Lopez, *Crossing Open Country* (New York: Vintage, 1989); John McPhee, *The Pine Barrens* (New York: Farrar, Straus, and Giroux, 1968).

20. Zube, *Landscapes*, 74, 77.

21. David Sopher, "The Landscape of Home," in *Interpretation of Ordinary Landscapes*, edited by Meinig, 144.

22. Yi-Fu Tuan, *Topophilia: A Study of Environmental Perception, Attitudes, and Values* (Englewood Cliffs, N.J.: Prentice-Hall, 1974); Yi-Fu Tuan, *Landscapes of Fear* (New York: Pantheon, 1979); Yi-Fu Tuan, *Dominance and Affection* (New Haven: Yale Univ. Press, 1984); Sopher, "The Landscape of Home," 144.

23. Zube, *Landscapes*, 116–31.

24. Ibid., 119.

25. Ibid., 119–23.

26. Ibid., 116, 131.

27. J. B. Jackson, "The Order of a Landscape," in *Interpretation of Ordinary Landscapes*, edited by Meinig, 153.

28. Meinig, "Reading the Landscape," 229.

29. Yi-Fu Tuan, "Rootedness Versus Sense of Place," *Landscape* 24 (1980): 3.

30. Meinig, *Interpretation of Ordinary Landscapes*, 164–92.

31. Ibid., 165, 167.

A Brief Meditation on Censorship

1. For example, see S. Alder and J. Brenner, "Gender and Space: Lesbians and Gay Men in the City," *International Journal of Urban and Regional Research* 16 (1992): 186–90; T. Geltmaker, "The Queer Nation Acts Up: Health Care, Politics and Sexual Diversity," *Environment and Planning D: Society and Space* 10 (1992): 609–50; P. Hindle, "Gay Communities and Gay Space in the City," in *The Margins of the City: Gay Men's Urban Lives*, edited by S. Whittle (Aldershot, England: Ashgate, 1994), 45–59; and P. Hubbard, "Desire/Disgust: Mapping the Moral Contours of Heterosexuality," *Progress in Human Geography* 24 (2000): 191–217.

In Revolutionary Mode

1. Richard Symanski and Nancy Burley, "Geography and Natural Selection—Revisited," Syracuse Univ., Discussion Paper Series, no. 25, 1976.

2. Richard Peet, ed., *Radical Geography: Alternative Viewpoints on Contemporary Social Issues* (Chicago: Maaroufa Press, 1977), v. There can be no doubt that Peet and Blaut found saviors and a lot more in the likes of Marx, Lenin, Stalin, and Kropotkin, among others. See Richard Peet, ed., *International Capitalism and Industrial Restructuring: A Critical Analysis* (Boston: Allen and Unwin, 1987); Richard Peet and Elaine Hartwick, *Theories of Development* (New York: Guilford Press, 1999); James M. Blaut, *The Colonizer's Model of the World: Geographical Diffusionism and Eurocentric History*

(New York: Guilford Press, 1993); and James M. Blaut, *Eight Eurocentric Historians* (New York: Guilford Press, 2000).

3. P. A. Kropotkin, *Ethics, Origin, and Development,* (New York: Dial Press, 1924), 5.

An Anatomy of Academic Censorship

1. As late as 1994, California, my home state, held the ignominious distinction of topping the nation in attempts to ban books in public schools. In that year, according to an annual report by People for the American Way, there were 43 attempted cases (9 of which were successful) in the state, and a total of 462 attempts to ban books and stop the showing of films and plays nationwide. Not only did censors go after perennial favorites such as John Steinbeck's *Of Mice and Men* and Mark Twain's *Adventures of Huckleberry Finn,* but in Modesto, California, *Webster's Ninth New Collegiate Dictionary* was attacked by parents because it contains "dirty words."

2. William H. Honan, "An Unexpected Debate on 'Unexpected' Content in Classroom Materials," *New York Times,* Oct. 5, 1994; "Professor Strikes a Blow for Academic Freedom," *New York Times,* Oct. 12, 1994. See H. Kramer and R. Kimball, eds., *Against the Grain: The New Criterion of Art and Intellect at the End of the Twentieth Century* (Chicago: Ivan R. Dee, 1995).

3. *Keyishian v. Board of Regents of New York,* 385 U.S. 589 (1967).

4. Coyote was an activist prostitute organization founded by Margo St. James in San Francisco. James was a former streetwalker who left the streets and actively campaigned for the legal rights of prostitutes. Coyote became the model for similar organizations in a number of other American cities, including Seattle and Miami.

5. Shortly after its publication, Butterworths informed me that the cost of editing and publishing the book, for a hardcover run of two thousand copies, was twenty-five thousand dollars. I spent approximately ten thousand dollars of my own money: two thousand dollars for the production of maps, and another eight thousand or so for field— and library work. I had no funding, academic or otherwise, for the book.

6. All letters from which quotations are taken are in my possession.

7. Symanski, "Prostitution in Nevada."

8. Gay Talese, *Thy Neighbor's Wife* (Garden City, N.Y.: Doubleday, 1980).

9. Symanski, "Prostitution in Nevada"; Symanski, "Manipulation of Ordinary Language."

10. See P. Jackson, *Maps of Meaning: An Introduction to Cultural Geography* (London: Unwin Hyman, 1989).

11. Richard Symanski, "The Responsibility of Richard Morrill," *Annals of the Association of American Geographers* 75 (1985): 136–38; Richard Morrill, "Comment in Reply to Richard Symanski, 'The Responsibility of Richard Morrill,' " *Annals of the Association of American Geographers* 75 (1985): 139; Richard Morrill, "The Responsibility of Geography," *Annals of the Association of American Geographers* 74 (1984): 2.

Index